Rebooting India

NANDAN NILEKANI
and VIRAL SHAH

Rebooting India

Research by Swapnika Ramu
Illustrations by Aparna Ranjan

ALLEN LANE
an imprint of
PENGUIN BOOKS

ALLEN LANE

UK | USA | Canada | Ireland | Australia
India | New Zealand | South Africa

Allen Lane is part of the Penguin Random House group of companies whose addresses can be found at global.penguinrandomhouse.com.

First published in India by Penguin Books India 2015
First published in Great Britain by Allen Lane 2016
001

Printed in Great Britain by Clays Ltd, St Ives plc

A CIP catalogue record for this book is available from the British Library

ISBN: 978-0-241-00392-3

www.greenpenguin.co.uk

This book is dedicated to everyone who has worked on Aadhaar.

We especially wish to acknowledge the contribution of the thousands of Aadhaar operators who work tirelessly to make the vision of Aadhaar a reality.

We owe our gratitude to Samar Halarnkar for his invaluable feedback that has made this book significantly more readable and interesting.

REBOOTING
INDIA

INTRODUCTION

> It must be considered that there is nothing more difficult to carry out, nor more doubtful of success, nor more dangerous to handle, than to initiate a new order of things.
>
> —Niccolò Machiavelli, *The Prince* (1532)

SEVENTY-YEAR-OLD BASUDEB PAHAN lives in a densely forested, remote area of Jharkhand. In order to receive his old-age pension from the Government of India, Pahan had to journey fifteen kilometres through hills and jungles to reach Ramgarh, the nearest settlement with a bank branch. And this was only half the story. To collect the 400 rupees a month owed to him, Pahan had to spend hours standing in line. Sometimes he needed to come back the next day. Factoring in the cost of travel and food, Pahan was spending over 12 per cent of his pension before he even received it. To add to his woes, he often had to wait two or three months for his payment to be processed. Pahan, the local government, and indeed the entire pension disbursement system were stuck in a time warp.

Then, in 2011, Pahan found himself transported from a dusty backwater of history into the forefront of India's technological revolution. He achieved this feat by walking a short distance to the local panchayat office in his village and using a device called a microATM, under the supervision of the local business correspondent appointed by a bank. The microATM, a handheld wireless device, required only an active mobile data connection in order to function. The business correspondent entered Pahan's twelve-digit Aadhaar number into the device. Pahan pressed his fingers on an attached fingerprint reader and, seconds later, the business correspondent was handing Pahan his money, just as a bank teller might.[1]

A smooth and seamless customer experience like Pahan's is usually typical of India's private sector, not its government, and serves as an illustration of a new, 'citizen first' way of thinking that must become the norm in our administration.

Pahan's fundamental problem—lack of access to banking facilities—might have been solved if a bank decided to build a branch in his village, an uncertain prospect. Instead of incremental change—building a banking network branch by branch and village by village—technology made it possible to deliver a bold solution overnight for Pahan, bringing banking right to his doorstep. We believe that technology holds the potential to completely redefine the relationship between the citizens and the state. What if a million banking correspondents equipped with smartphones and biometric readers could deliver banking services to 1.2 billion Indians? What if every citizen could use mobile banking and make cashless payments? Moving beyond the financial sector, can we use technology to solve some of India's most pressing problems—to improve standards in healthcare and education, to cut wastage in government spending and increase revenue, to make tax collection citizen-friendly, to streamline our courts, and to eliminate corruption?

These and other questions form the focus of our book, and in every chapter we propose some 'big ideas' that can radically redesign existing systems, and in the process save the government an estimated $15B (Rs 1 trillion) annually. Conservative back-of-the envelope calculations find those savings to be equivalent to 1 per cent of our GDP—enough for two Golden Quadrilateral road systems across the country every year,[2] or to send 200 Mangalyaan missions to Mars annually.[3] One year's savings are also sufficient to provide minimal health insurance for every family in the country for three years.[4]

Software is eating the world

In 2011, Marc Andreessen, the co-founder of Netscape, famously proclaimed, 'Software is eating the world. More and more major businesses and industries are being run on software and delivered as online services—from movies to agriculture to national defence. Many of the winners are Silicon Valley-style entrepreneurial technology companies that are invading and overturning established industry structures.'[5] For the most part, these businesses and services have come into existence only in the last decade, but have transformed our social landscape to the point that we now wonder how we ever got by without them. Until the 1990s, having a telephone connection was a prized rarity that took years to obtain; today, India's mobile telecommunication network is the second largest in the world, with over 900 million users. Call rates used to be 16 rupees a minute;[6] at the current rate of one paisa per second, our call rates are among the lowest in the world. India now boasts of the world's third largest internet user base, with over 190 million users, many of whom are using smartphones to get online and buy things; as much as 40 per cent of all e-commerce transactions in India are now conducted via mobile phones, bypassing computers altogether.[7]

Whether it's to watch a movie, a cricket match or a play, many of us don't throng box office windows any more, but buy our tickets from websites like BookMyShow. We book train tickets through the website of IRCTC (Indian Railway Catering and Tourism Corporation), look up flights on MakeMyTrip or Yatra, and use RedBus if we're to travel by bus. We book our cabs online through services like Ola or Uber, tracking their location through GPS and paying the fare electronically without having to hunt through our pockets for change.

You can now order groceries from your local kirana store on WhatsApp and have them delivered to your home, a business model that many small entrepreneurs are following. BigBasket and Grofers work on the same principle, but on a much larger scale. If you can buy fruits and vegetables over the internet, why can't a farmer order fertilizers and seeds online, and have them delivered to his door? Why can't a poor family order cooking fuel using a smartphone? Why can't crop insurance funds be automatically transferred to a farmer's bank account as soon as a drought is declared?

You can now take personalized courses on a variety of subjects through Skype, and students across India now have access to classes from some of the world's best universities through online platforms like Coursera, Udacity and edX. Is it possible to achieve the same outcome for every student in every government school in India? You can now find a doctor through your smartphone thanks to businesses such as Practo, which allow users to rate medical practitioners based on their quality of service. How can government primary healthcare centres and hospitals deploy technology in a similar way?

The gap between dreams and reality

Rayappa Pitkekar is a forty-eight-year-old leatherwork contractor in Mumbai's Dharavi locality; its status as one of the world's largest slums belies the fact that its flourishing informal economy boasts a turnover of over $460M (Rs 30 billion) annually,[8] thanks to the enterprising spirit of Pitkekar and his ilk. The son of a garbage picker who earned 2 cents (Rs 1.25) a day digging through trash to salvage plastic for reuse by the local plastic industry, Pitkekar studied by day and then worked two jobs, earning 8 cents (Rs 5) for every 100 newspapers he sold in local trains in the evening and then working in a hotel at night. His leather goods business now has an annual turnover that runs into lakhs of rupees; his daughter Priyanka's dream is to 'become what Kalpana Chawla was [an astronaut]', while his son Kunal wants to be a doctor.[9] Within three generations, Pitkekar's family, rooted in the hard-scrabble life of a Mumbai slum, is now literally aspiring for the moon.

For the first few decades of our country's existence as an independent nation, we turned our gaze inwards, preoccupied with building the dams, steel plants, universities and cities that became the first temples of modern India. As our country's infrastructure began to take shape, we tackled such concerns as poverty reduction, while ushering in the Green Revolution and economic liberalization, the last of which closed the book on the isolationist and protectionist policies of an earlier era, and recognized the need for our country to take its rightful place on the global economic stage. The winds of liberalization brought in a new mindset for India's people; no longer content to plough the same furrow that their forefathers had before them, their ambitions began to grow. They wanted a system that would support their efforts to climb out of poverty and ensure a brighter future for themselves and their children.

India is sitting on a demographic dividend and is expected to become the world's youngest country by 2020, with 64 per cent of its population, roughly 800 million people, of working age.[10] That is 800 million knocks on the ceiling with a list of demands that include education, employment, good health, better infrastructure, efficient governance and a corruption-free society. The economy that is supposed to sustain the weight of these demands has been growing in single digits, around 9 per cent a year in the best of times—a flimsy scaffold on which to construct dreams of a better life. How do we build a foundation strong enough to nurture these dreams and bring them to fruition?

The idea of government as an enabler of people's aspirations necessarily implies a radical rethink of the relationship between our government and its people, one that still seems stuck in a bygone era in its reluctance to embrace technology's transformative power. Despite all the advances in technology—the 'dotcom revolution' that India embraced so enthusiastically, the new business models that are springing up every day, the way India's online footprint keeps growing—until recently, Basudeb Pahan and millions of others like him across India still had to line up at a government office or a bank branch, wait for hours, return multiple times, and pay for transportation and opportunity costs to receive a payment that only arrived on an errant schedule. Whether it was getting a pension, a government subsidy or affordable food from a ration shop, the process remained hopelessly complex and time-consuming, mired in bureaucracy and local interests, and imposing a heavy financial toll on the poor.

Even for an urban, middle-class Indian, dealing with the government is cumbersome. Getting a driving licence, a passport, a birth or death certificate is a long-drawn-out affair, necessitating multiple trips to the relevant government office. Citizen service centres like Bangalore One and the recent initiative by the Government of Karnataka to create the nation's first mobile governance platform are laudable attempts to make these interactions less fraught, but are so far restricted to small segments of the population.

Ask any small business owner in India what his or her biggest complaint is, and chances are that 'compliance with local laws' will rank high on the list. Merely starting a new business in India takes weeks; most of this time is spent in completing the required paperwork and legal formalities. Whether it's paying taxes or negotiating complex labour-law requirements, we haven't yet built a truly entrepreneur-friendly environment, where anyone with a bright idea and some capital can easily start a business. Consider the fact that out of 189 economies surveyed by the World Bank to compile the 'Ease of Doing Business' index, India ranks at a deplorable 142nd, significantly lower than all its BRICS (Brazil, Russia, India, China, South Africa) counterparts.[11] All in all, the discomfort of these interactions has been pithily summed up by the authors John Micklethwait and Adrian Wooldridge as governments attempting to 'govern the world of Google and Facebook with a quill pen and an abacus'.[12] How can government use technology to deliver a customer-service experience comparable to that on offer in the private sector today?

Since Independence, we have built many great institutions which have withstood the test of time—our Constitution, universal suffrage, the parliamentary system, our courts, the civil services, the federal structure and the separation of powers, among many others. For the world's largest and most diverse democracy, born in abject poverty, scrambling for every possible resource, this is no mean feat. The checks and balances governing our democracy have largely worked. However, we as a nation seem to have fulfilled Babasaheb Ambedkar's oddly prophetic prediction: 'In political life we will have equality and in social and economic life we will have inequality.'[13] The fruits of the nation's development have not been equitably distributed across our population. The stark divides between the haves and the have-nots continue to exist, and are getting sharper than ever. Our formal sector—the section of society that holds government-issued IDs (in a pre-Aadhaar era), pays its taxes, takes out loans—is a thin layer on top of a vast, self-organized informal sector, which remains largely outside the purview of the government, struggling to claw its way up the ladder of economic prosperity.

We expect our elected politicians to provide a springboard for our aspirations. We expect them to create an environment where children go to school for high-quality education, where good healthcare is easily accessible to everyone, where entrepreneurs thrive and create enough well-paying jobs for every able person. We expect our government to be the provider of the last resort—to build social safety nets for those unfortunate enough to have fallen through the cracks of society. We no longer want to be supplicants to a '*mai–baap sarkar*', uncomplainingly accepting the poor quality services it doles out; we want a government that leverages the strength of our institutions, the power of our markets and the industriousness of our people to build high-quality solutions at speed and scale. The politics of India is the politics of the past—marked by caste inequalities, religious conflicts and reservations. The politics of the future is the politics of meeting a billion aspirations, the weight of which will crush anyone who fails to deliver.

Betting big on technology: The trend is your friend

Whenever Sanjay Sahni, a school dropout working as an electrician in New Delhi returned to his home village of Ratnauli in Bihar's Muzaffarpur district, he would be besieged by complaints from villagers that they weren't getting their dues under the government's MGNREGA (Mahatma Gandhi National Rural Employment Generation Act) scheme, meant to provide guaranteed employment and wages to rural residents. Sahni used to store his tools in a New Delhi cybercafe; knowing nothing about computers, one day he impulsively typed 'NREGA Bihar' into Google, and found the official list of villagers who were supposedly beneficiaries. As it turned out,

1 SCALE

Solutions that handle millions of people and billions of transactions

2 SPEED

Solutions that can be developed in months and years, not decades

3 COST

Solutions that decrease process and service costs

4 ENFORCEABILITY

Solutions that can be monitored in real time

5 DIVERSITY

Solutions that work as platforms to foster innovation

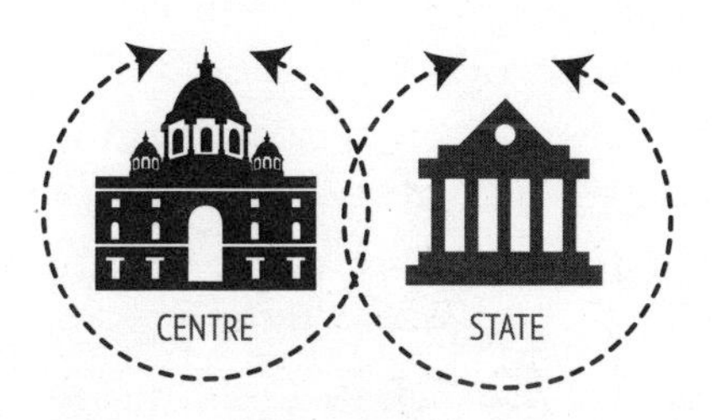

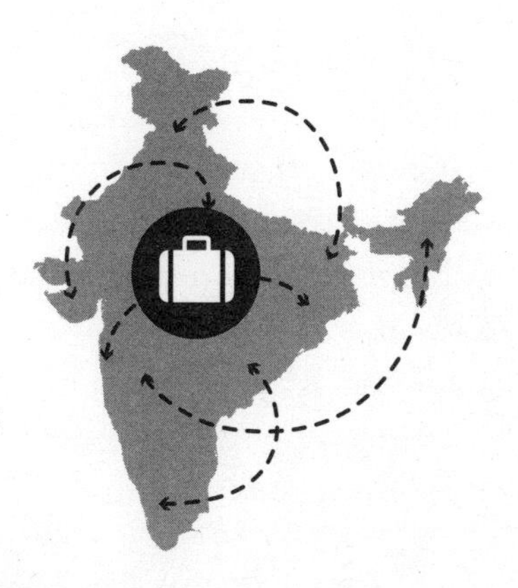

6 AUTONOMY

Solutions that allow government (central, state, local) and its agencies to function independently

7 MOBILITY

Solutions that are accessible anywhere in the country

8 INTEGRATION

Solutions that incorporate the best components across the public–private spectrum

9 COLLABORATION

Solutions that share information and develop partnerships across government

10 INCLUSION

Solutions that lower entry barriers and widen access for all

many of those listed hadn't been given any work, and those who had didn't get paid in full. The money was being diverted to contractors, village headmen and government officials instead. Armed with this knowledge, Sahni successfully fought for his fellow villagers' right to employment and a fair wage.[14] The government didn't have to mount a sophisticated anti-corruption scheme; all they did was to digitize information and make it easily accessible, and the public did the rest. This isn't just the archetypal story of the common man triumphing against the system; it marks a far more profound shift in the balance of power between the government and the people, one that is only possible because of technology.

Indian society is vastly complex, made up of a million smaller Indias—rich, poor, urban, rural, multi-ethnic, multilingual. A truly democratic society is one in which every citizen of these million smaller Indias is included in the mainstream, able to benefit from government services and participate in the country's development. We need to fix our country's problems at great speed, at scale, with high quality, while providing solutions that are easy to access, independent of geography, and low-cost. Technology, the great leveller, is our only hope of meeting these goals.

When it comes to speed and scale, consider that Facebook successfully manages a database of one billion users; at one point, it was adding 100 million users roughly every five months.[15] WhatsApp reached 900 million users much faster. Today, we can top up our mobile phones for as little as five rupees, the most common amount for a top-up in the country. As a result, more people can own a mobile and make calls. We leapfrogged past landlines, and went from no phones to almost everyone having a phone in the span of fifteen years. Aadhaar will cover the entire country in less than ten years.

The ability of technology to deliver low-cost solutions is especially important in a developing country like ours, and this approach has already proven its worth many times over. For example, withdrawing money from a bank used to take anywhere from ten minutes to an hour, and cost the bank an estimated Rs 50 in processing fees; with the advent of ATMs, the same transaction can be completed under five minutes at a cost of only 23 cents (Rs 15). The money saved in transaction costs is enough to pay for the set-up and maintenance of ATM machines. The National Stock Exchange (NSE) and the National Securities Depository Limited (NSDL) have helped us make the switch to a fully electronic online stock-trading system, bringing down the costs of trading and storing securities to a fraction of what they were a decade ago. In the five years from 2000 to 2005, the cost per transaction at NSDL crashed from 35 cents (Rs 23) to 8 cents (Rs 5), an even sharper decline if adjusted for inflation. It now stands at 7 cents (Rs 4.5).[16]

Another attribute that makes technology-driven solutions so attractive is their ability to transcend the limits of geography. Today, we are a nation on the move. In the last two decades, India's urban population has exploded, thanks to a steady influx of people migrating from villages and small towns in search of education, employment and a better way of life—at last count, India's migrants numbered over 300 million.[17] Technology has managed to keep pace with this exponential rate of migration—for example, a mobile phone connection registered in an Andhra Pradesh village works smoothly in Bangalore. Your bank account may have been set up in Mumbai, but you can check your bank balance anywhere in the country.

In comparison to this fluidity, government services remain highly rigid, inescapably tethered to a physical location. Nandan contested the 2014 general elections from the Bangalore South constituency; in the course of his election campaign, we met many people who had recently moved to the city and couldn't vote because their voter IDs hadn't been transferred from their original location. Migrants who move to cities find that their village ration cards won't allow them to obtain subsidized foodstuffs from city ration shops. Pension payments can only be collected from specific bank branches. The private sector has successfully employed technology to deliver services that transcend geographical barriers; it is time the government followed suit.

The speed at which technology can evolve is the subject of Moore's Law, the famous prediction by Intel co-founder Gordon Moore that computing speed would double every eighteen months—a vision

DESIGN OF GOVERNMENT SYSTEMS

High entry barrier based on eligibility, proof, documents

Free ride once inside, even if the system detects fraud

Single point of entry

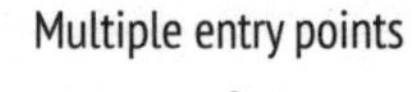

THE NEW MODEL

Multiple entry points

Robust technology to weed out fraud and encourage compliance

MAXIMIZING CITIZEN CONVENIENCE

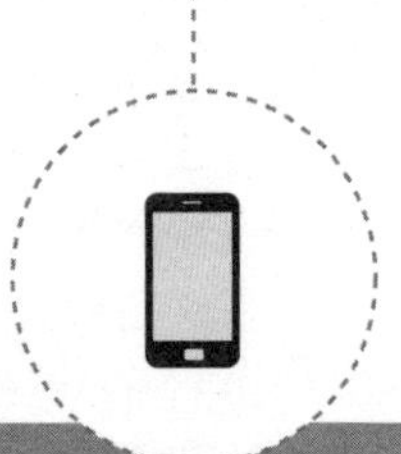

TRANSPARENCY AND EASE OF ENTRY

INCLUSION AS THE DRIVER

of exponential growth that has largely been realized. The average smartphone today is more powerful than the computers that put the Apollo astronauts on the moon. Five years from today, they will be more powerful than the supercomputers of the last decade. Over the last few decades, the internet has become a thousand times faster while computer data storage has become a thousand times cheaper. These developments have resulted in the emergence of cloud computing, where you can rent high-powered computers from Amazon and Microsoft for twenty rupees an hour—the cost of a few cups of tea. Today, YouTube probably generates more data through uploading and streaming videos in a few days than the medical industry does in an entire year. While they're getting cheaper, faster and smarter, computers are also getting tinier and more ubiquitous, embedded in everything from shoes to clothes, eyeglasses to robots. These are the key attributes of technology we need to consider when designing solutions to India's many challenges.

The A-Team: How 101 people can fix all of India's problems

Within the past two decades, popular culture has acquired a new, reliable trope—programmers hacking away in a small garage or dorm room, fuelled by soda and junk food, creating a start-up that eventually takes over the world and rakes in billions. It's a testament to the power of the idea that a small, committed team can achieve what many large organizations cannot. When Google paid over a billion dollars for YouTube, it was acquiring a company that was one of the largest consumers of bandwidth on the internet, while it employed only around forty people. When the popular photo-sharing app Instagram was bought for a billion dollars by Facebook, it had only twelve employees. What if the same start-up model could be applied to solve the problems of governance as well?

As alien as it may sound, India has actually nurtured a proud tradition of start-ups within government ever since Independence. The very first of these was the Atomic Energy Commission, founded by the physicist and bon vivant Homi Bhabha, as atypical a 'sarkari babu' as one could hope to find. With a small, dedicated team of colleagues, he built India's indigenous atomic energy programme, meant to meet the country's power needs and bolster national security. Another outstanding example is that of Vikram Sarabhai; the organization he established, the Indian Space Research Organization (ISRO), continues to win plaudits to the present day, with successes like Chandrayaan and Mangalyaan under its belt. In her book *Vikram Sarabhai: A Life,* Amrita Shah writes:

> By 1970, his list (of space applications) had expanded to include agriculture, forestry, oceanography, geology, mineral prospecting and cartography . . . This was Vikram's dream: linking technology with development, serving the needs of the masses while nurturing a highly sophisticated work culture and scientific abilities. One of his favourite phrases was 'leapfrogging'. It referred to his great faith, along with Bhabha and Nehru, in the ability of technology to enable developing countries to circumvent the long, arduous processes followed by the Western world.[18]

India's self-sufficiency in foodgrains and milk are also the results of start-ups: the Green Revolution, powered by scientists like Norman Borlaug and bureaucrats like M.S. Swaminathan, and supported administratively by the Indira Gandhi government, is responsible for the former, while Operation Flood, spearheaded by Verghese Kurien, made India the largest milk producer in the world. In both cases, seemingly intractable problems were tackled by small teams that were given strong administrative support—the results speak for themselves. Fast-forwarding to our own experience, a dedicated group of bureaucrats and technical specialists working in tandem were able to build and deliver on the promise of the Aadhaar programme—handing out a unique identity to every one of India's 1.2 billion residents. Around the same time as Aadhaar, the Bharat Broadband project was also launched, aimed at connecting 250,000 villages with fibre-optic networks. Aadhaar was deliberately designed from the very start for scale and speed, and over 900 million people now have an Aadhaar number; Bharat Broadband's design inefficiencies, on the other hand, led to a gridlock, and the project is still floundering. Getting the design right in the beginning is key.

We propose that a team of 100 carefully selected individuals can fix all the major problems that ail India. How would such a system work? Let's say the prime minister identifies ten grand challenges that India faces. Each idea can be the nucleus of a 'government start-up'. Ten enterprising leaders are given charge of each of these problems. They, in turn, form ten-member teams of the best brains within government and domain specialists from outside government to apply out-of-the-box thinking that can deliver innovative solutions. We give examples of such potential solutions throughout this book.

Any new government project should be treated, in essence, like a start-up that needs to stake a claim for itself. The officials in charge of such a project need to display a considerable amount of entrepreneurial savvy. A true entrepreneur will figure out all the government processes and follow them to the letter. He will navigate the byways of the bureaucracy, keep his multiple masters happy, get his project mentioned in every important speech and every government document of relevance, get his bills tabled in Parliament and enacted as law, secure his budgets, cooperate with investigating agencies, respond to court orders, answer Parliament questions, tirelessly provide information sought in RTI requests, build general consensus with multiple interest groups within government as well as citizen groups outside, find allies who will support him when under attack, and do all this while staying focused on hiring the best team and building an organization that is dedicated towards achieving a well-defined goal.

The standard manual of 'business as usual' must be thrown out of the window. A bureaucracy consisting of officials who are experts in administrative procedure alone will not be able to cope with the kind of large, complex projects that our government needs to tackle. While there are many talented, hard-working and honest bureaucrats, we must recognize the shortcomings of this system as well: a hard-coded hierarchy that places a premium on seniority; territorial battles; a bias for complexity; the shortness of tenure that creates short-term vision;

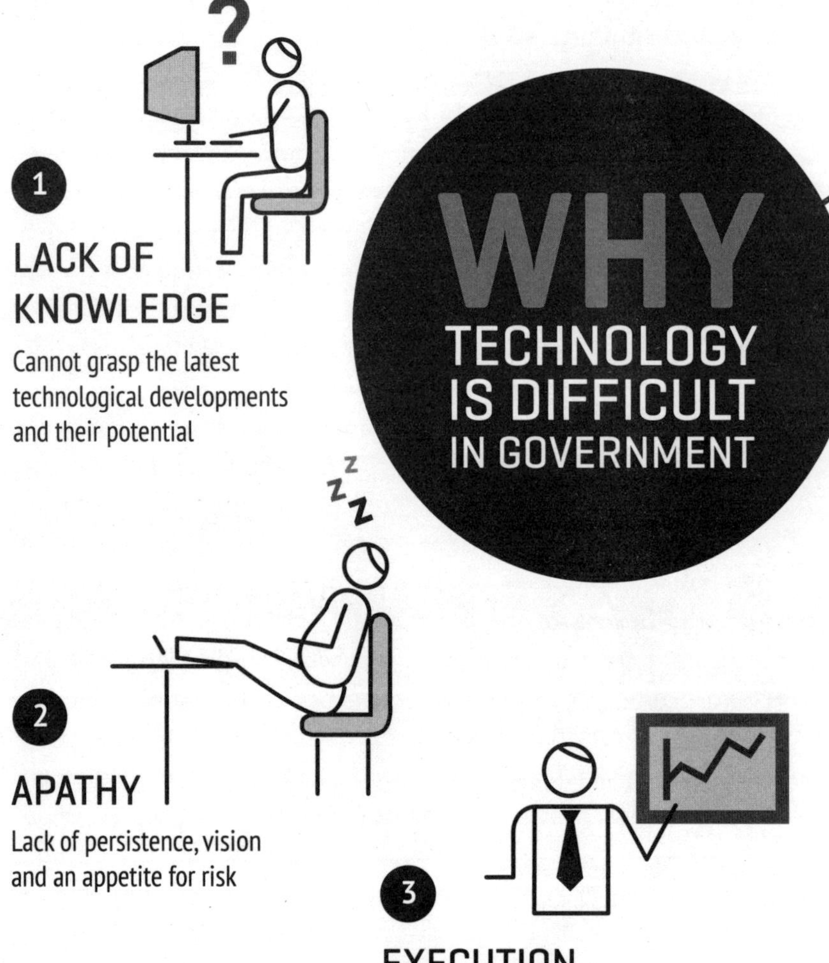
WHY
TECHNOLOGY
IS DIFFICULT
IN GOVERNMENT
1
LACK OF
KNOWLEDGE
Cannot grasp the latest
technological developments
and their potential
2
APATHY
Lack of persistence, vision
and an appetite for risk
3
EXECUTION
Lack of skill to manage
complex technological projects

4
POWER SHIFT
TO AUTOMATION
Removes opportunities for
corruption and rent-seeking

POWER SHIFT TO NEW PLAYERS

Resistance to changing traditional models of operation

COORDINATION

System favours territory and hierarchy over collaboration

7

PRIVACY & SECURITY

The spectre of data vulnerability and 'big brother'

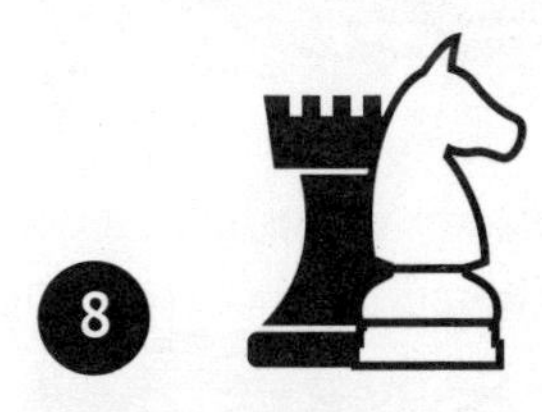

8

TACTICS OVER STRATEGY

Short-term vision that prioritizes patronage over the public good

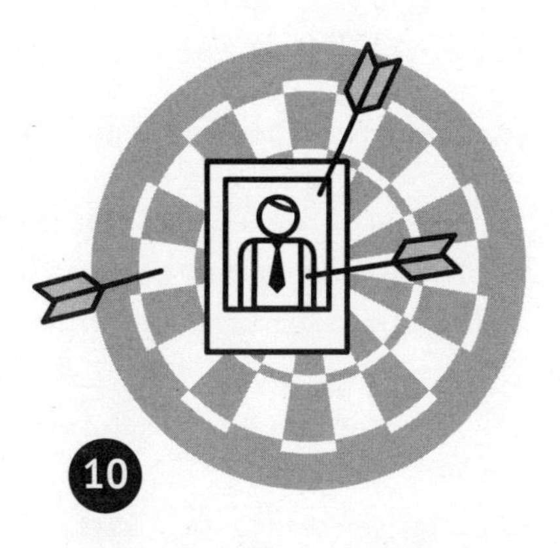

9

FEAR OF CHANGE

Comfort with the status quo

10

NEGATIVE COALITION

Ganging up to retain power and fight transformation

the idea that power lies in commanding the largest number of people with the largest possible budget; and inter-service rivalry.

Why is government so byzantine? The Federalist Papers, published in 1787–88, discuss at length the design decisions made during the drawing up of the American Constitution. In Federalist Number 51, James Madison, the fourth President of the United States, expounds upon the controls that must be necessarily placed on the government:

> But what is government itself, but the greatest of all reflections on human nature? If men were angels, no government would be necessary. If angels were to govern men, neither external nor internal controls on government would be necessary. In framing a government which is to be administered by men over men, the great difficulty lies in this: you must first enable the government to control the governed; and in the next place oblige it to control itself.[19]

This is the challenge that lies before us. How do we usher a new class of people into government without weakening the structure of governance, while simultaneously harnessing the energy of the start-up culture? We believe that our experiences with Aadhaar will highlight some parts of the solution.

A road map for rebooting India

Shankar Maruwada, former head of demand generation at the Unique Identification Authority of India (UIDAI), who has spent his career thinking about marketing in India and the Indian customer's psyche, tells us that his team did a lot of ground research and spent a good six months travelling across the country before Aadhaar was officially launched. The team built a picture of the real India from both rural and urban perspectives. He says, 'We found that India's biggest brand is the "*sher chhaap*"—the symbol of the government.' People really do believe that the government has their best interests at heart, and trust it implicitly, even if the frontline workers and officials they interact with are corrupt or inefficient. A tremendous reservoir of goodwill and faith has been built despite our creaking and outmoded public service systems. If the government were to embrace technology and radically redefine the way it interacts with its subjects, imagine the kind of positive energy it could generate, an energy that could power our country to new heights of achievement and prosperity.

This book is fundamentally about the ways in which technology can be employed to transform our government and create a series of new, citizen-friendly public institutions, based on our experiences building the Aadhaar project, the world's largest social identity programme. The economist Lant Pritchett, a friend of Nandan's, calls India a 'flailing state', one where the head is functioning but has lost the connect with its limbs; a condition in which 'the everyday actions of the field-level agents of the state—policemen, engineers, teachers, health workers—are increasingly beyond the control of the administration at the national or state level'.[20] We are a 'booming economy and democracy with world-class elite institutions' while simultaneously contending with 'chaotic conditions in service provision of even the most rudimentary types'. Our country cannot become a global powerhouse unless we resolve the contradictions and bridge the gaps that are distorting our society.

The first part of our book focuses on the idea of Aadhaar as a platform upon which to build a new class of services. These include an electronic payments network, a redesigned social security system and the electronic Know Your Customer (e-KYC) service. We then examine the idea of technology as a tool for institution-building, focusing upon the Goods and Services Tax (GST) and Electronic Toll Collection (ETC) as the building blocks of a single market. Later, we take a look at the electoral process in India, both in terms of the way elections are run and contested—in the latter context, we look at how technology has given birth to the concept of the party as a platform. Finally, we turn to the bigger picture, looking at how technology can be used to answer some of India's most pressing governance concerns such as justice, health and education.

President Pranab Mukherjee provided an excellent analogy to

explain the impact of technology on government. During Mukherjee's tenure as finance minister, Nandan met him to submit a report on the issue of subsidy reform, and emphasized the importance of technology platforms in solving some of the most complex problems that India faces. Mukherjee immediately understood the point, turned to the various other ministers and secretaries, and explained, 'See, it is just like a railway platform. Different trains pull up at a railway platform, each with a different destination, and people get on and off depending on where they are headed. In the same way, the technology platform is a central location where various state governments, institutions, and citizens can gather. All government services and schemes are offered on the same platform, and citizens can enrol for all eligible services in one place.'

The challenge that lies before us is to create the platform that allows every one of India's 1.2 billion people to get on a train headed for the destination of their choice.

1

AADHAAR: FROM ZERO TO A BILLION IN FIVE YEARS

Poor people are bonsai people. There is nothing wrong with their seeds. Only society never gave them a base to grow on.

—Muhammad Yunus, *Creating a World without Poverty*

TEMBHLI VILLAGE IS a small, dusty hamlet in Maharashtra's Nandurbar district, close to the Gujarat border. Inhabited mostly by Bhil tribals and fifty kilometres from the nearest town, Tembhli did not have proper roads or electricity, and its residents either laboured for a pittance in the nearby fields or migrated to other states in search of employment. Then, in late 2010, Tembhli underwent a dramatic makeover. Roads were asphalted, fresh coats of paint were slapped onto buildings, water pumps were fixed and the village got its first-ever electricity connection. All this was in honour of the fact that Tembhli was chosen as the place where the Government of India would flag off the ambitious Aadhaar programme. In a function attended by the then prime minister Manmohan Singh and several other dignitaries—including Sonia Gandhi, president of the Indian National Congress—Ranjana Sonawane, a thirty-year-old housewife, officially became the first Indian resident to receive an Aadhaar number. The villagers hoped that their Aadhaar numbers would enable them to do the simple things—keep track of their government records, open a bank account, ensure that nobody was fraudulently claiming their benefit payments—that had been denied to them so far.

The sight of a beaming Sonawane posing for photographs holding the official card on which her unique ID number was printed wasn't particularly novel—we've seen similar pictures every time a new government scheme is launched. What was different was the power of that number clutched so proudly in Sonawane's fingers. Firstly, Aadhaar is the only universal identity available to all Indian residents, even those who do not possess any other form of identification, making them visible to the state for the first time. Secondly, Aadhaar is that rare government scheme that was not designed to address a single need. Instead, it is an open identity verification system that can be plugged into any application that requires an individual to prove who they are, whether they're enrolling for a rural job guarantee scheme or opening a bank account. No other identity scheme has used biometric identity verification—using a person's fingerprints and iris to validate their identity—on such a large scale. And finally, Sonawane was not required to present her official card every time she wanted to use her Aadhaar number—as long as she knew what her number was, her eyes and fingers were sufficient to declare to the world that she, and she alone, was Ranjana Sonawane of Tembhli.[1]

Ranjana Sonawane was the first of over 900 million Indian residents—over two-thirds of the 1.2 billion-strong population—who now have Aadhaar numbers.[2] For those of us in the audience that day in Tembhli—or on the stage, in Nandan's case—she was a symbol of the remarkable journey we had undertaken in the past fourteen months as part of the Unique Identification Authority of India, the organization set up to implement the Aadhaar scheme. Many talented individuals, from government agencies and private companies alike, made personal and professional sacrifices to become part of the programme. We'd set up shop in half-built offices and semi-furnished apartments, gulped down endless cups of coffee as we worked late into the night, and travelled the length and breadth of the country to meet chief ministers and CEOs alike, all because we were committed to an idea—the idea that every Indian deserves an identity, and a way to answer that most fundamental of questions: Who am I?

It took less than five years for Aadhaar to reach over two-thirds of the country; that's more people acquired in a shorter time than WhatsApp, the mobile messaging application, which was bought by Facebook for $19 billion. In those five years, we at the UIDAI built the entire framework of the programme from the ground up, and had to work with multiple government and private agencies to achieve our goals on time. And we did all of this entirely within the ambit of the Indian government, working within the same system that so many scathingly dismiss as sclerotic, old-fashioned and inefficient. We found people with vision, commitment and a passion for change wherever we looked. This, in our opinion, was one of the great victories of Aadhaar.

India's invisible millions

In 2010, a man named Farooq Alam signed up to receive an Aadhaar number in New Delhi.[3] He was a migrant labourer from Uttar Pradesh, with a wife and family he had left behind in his quest to try and make a living. He was not sure of his own age; like 59 per cent of children born in India,[4] his birth had never been registered, and he had no birth certificate. Without proof of his age, Ali wasn't able to obtain a voter ID or a driver's licence. In fact, Ali had no form of identity documents at all. After he completed the enrolment process, he was asked how things would change now that he possessed a document that proved who he was. 'Maybe some benefits,' he said, while his co-worker Gulzar Khan was far more blunt, 'I never had ID, so I never thought about what it would mean, but now that I have it, I think that at least people will know who I am if I'm killed in an accident.'

This is what it was like to be one of the millions of Indians who possessed no identity documents, who had no way of proving who they were, and whose birth, life and death were completely invisible to government authorities. This lack of identity manifested itself in a thousand missed opportunities, both large and small. These 'invisible' Indians could not be employed in the formal labour market without identification, consigning them to physically demanding, poorly-paid jobs in the country's vast unorganized labour sector. They could not register as voters, and so had no say in choosing their leaders. They could not legally sign up for any of the government welfare schemes meant to benefit the poor. They couldn't drive a car, travel out of the country, get a SIM card for a mobile connection, open a bank account or provide any documentation if the police stopped them on the street. They remained mired in poverty, unable to find any way out of their circumstances.

The barriers to entry for obtaining any form of identification were insurmountably high for most of India's poor. Every major form of ID in India required you to already be in possession of at least one other form of ID, be it a birth certificate, a voter ID, a PAN card or a ration card. The Kafkaesque complexity of merely trying to prove who you are meant that the underprivileged who are most in need of the government's social safety nets simply fell through the cracks.

To make things even more complicated, not all these forms of ID were available to everyone—you need to be above eighteen to have a voter ID or a driver's licence, for example. And none of these documents was universally accepted as a proof of identity. This meant that every time you wanted to carry out any transaction that required identity verification—say, opening a bank account—you would have to painstakingly assemble an entire sheaf of documents that were required by the bank for this process, and fill out a set of forms that were only applicable to that bank. The whole rigmarole was tedious, costly and time-consuming, again placing these services out of reach of those who needed them the most.

Identity verification was equally painful from the bank's end, requiring a great deal of effort to be expended. Worse still, whether it was banking or any other service, all this data would end up in a silo. In other words, the data you submitted when applying for a driver's licence was not accessible to the passport office if you applied for a passport—you had to go through the whole process all over again. Such opaque, inaccessible data systems meant that it made it hard to find and remove errors, increasing the potential for fraud to go unchecked.

Fake beneficiaries and ghost workers existed in the thousands, created to illegally withdraw benefits, while the intended beneficiaries were cheated out of their dues and remained invisible to the system.

If the fruits of India's economic progress were to truly reach every resident of the country, our invisible millions would have to emerge from the darkness to which they had been relegated and take their rightful place as part of the formal financial sector. At the same time, government benefit schemes were haemorrhaging billions of rupees—for instance, nearly $4.6B (Rs 300 billion) disappeared every year from the public distribution system for wheat, rice and kerosene. These leakages had to be staunched if such aid programmes were to become truly effective. It was against this background that the Government of India launched a scheme to issue a Unique Identification Number (branded as Aadhaar, which means 'foundation' in various Indian languages) to all 1.2 billion residents of India. As we explain in the following diagram, this number is meant to be universal—unlike other existing forms of identification—and can serve as the sole information required for identity verification in order to obtain a host of government services and benefits. By reducing the cost of identity verification and developing an inclusive platform for enrolment, the goal of the programme was to make every Indian, no matter how poor or marginalized, visible to the state.[5]

Many decades ago, the then prime minister Rajiv Gandhi had famously declared that for every rupee spent by the government, only 17 paise reached the intended recipient, with the rest lost to corruption. The increased transparency and accountability brought about by the Aadhaar programme will go a long way in plugging these leaks. The Aadhaar number can serve as the basis of a model in which benefits are directly transferred to the recipients; in the next few chapters, we explore multiple scenarios in which such a model is now being utilized. The direct benefit transfer system allows for better tracking of resources and greater transparency, eliminating the problem of ghosts and duplicate identities along the way. This system, while being simple and convenient enough for anyone to use, also provides the high levels of security that government regulations demand.

THE ISSUE OF **IDENTITY**

Most government-provided IDs are function-driven and crafted for an administrative purpose

PASSPORT
Only for citizens to travel outside the country

DRIVER'S LICENCE
Only for vehicle drivers 18 years and over. Limited validity

PAN CARD
Only for taxpayers

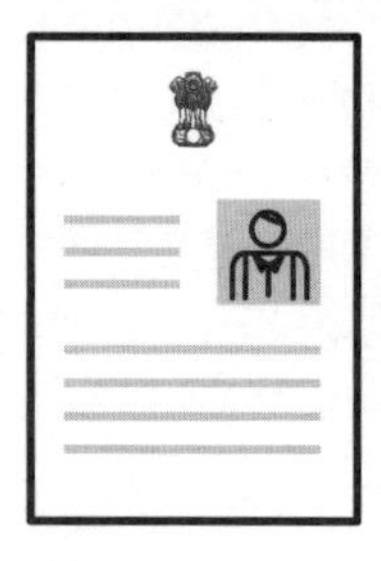

VOTER ID
Only for citizens 18 years and over

RATION CARD
Only for citizens who qualify for subsidies

BPL CARD
Only for citizens falling below a certain level of income

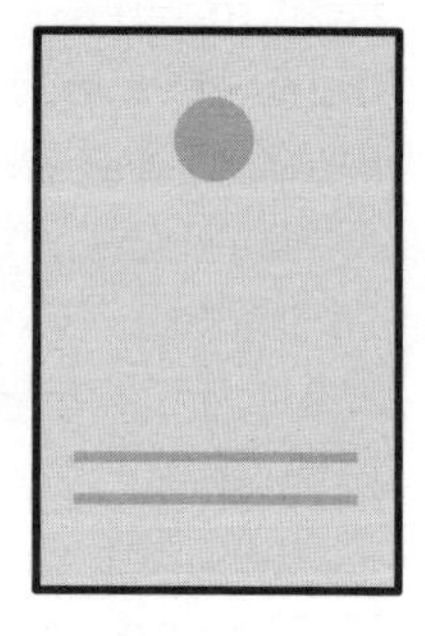

NREGA CARD
Only meant for rural employment scheme

AADHAAR

AADHAAR CARD
Primary function is to provide identification of the individual

- Universal • Digital
- Secure verification
- Multiple applications like e-KYC, Subsidies, Pensions
- Can be used as a financial address

Providing a universal identity and propelling the creation of a direct benefit transfer system were two of the immediate outcomes envisioned when the Aadhaar scheme was first proposed. Given the transformational power of technology, we anticipate that an individual should soon be able to use their Aadhaar number for an untold number of applications that require identity verification.

An unexpected journey

'Such a "national grid" would require, as a first and critical step, a unique and universal ID for each citizen. Creating a national register of citizens, assigning them a unique ID and linking them across a set of national databases, like the PAN and passport, can have far-reaching effects in delivering public services better and targeting services more accurately. Unique identification for each citizen also ensures a basic right—the right to "an acknowledged existence" in the country, without which much of a nation's poor can be nameless and ignored, and governments can draw a veil over large-scale poverty and destitution.'

Words from the UIDAI's mission statement? No. These words are taken from Nandan's earlier book, *Imagining India*.[6] As far back as 2008, Nandan had devoted serious thought to the question of a national ID programme. Around the same time, the government had taken its first steps towards implementing such a programme, an idea which had first been mooted in 2006. The UIDAI formally came into existence on 28 January 2009, and Nandan's involvement began almost immediately, despite the fact that he was not officially appointed chairman until later in the year.

In the summer of 2009, Viral was living in Santa Barbara, California, surfing, playing Ultimate Frisbee, nursing a serious burrito habit and working for a Boston-based start-up while also taking finance courses at his alma mater, the University of California, Santa Barbara, where he had completed his PhD in computer science and engineering. As part of his studies, he was given the chance to attend a conference in New Delhi organized by the National Council of Applied Economic Research (NCAER). It so happened that Nandan was also attending the conference; the two bumped into each other in the hallway, and introductions were made by the economist Ajay Shah, who told Viral, 'Nandan is working on something really important and you need to be a part of this.' That hallway encounter led to a fifteen-minute meeting in the lobby of a New Delhi hotel, where Nandan explained in detail what this 'something important' was. Viral returned to the US; his start-up was acquired by Microsoft, and deciding that he didn't want to be a cog in the corporate machine, he began scouting for other opportunities. One option was a postdoctoral fellowship in the mathematics department at the Massachusetts Institute of Technology (MIT), but this required a change in his visa status. In September that year, Viral found himself back home in Mumbai, doing the rounds of the US consulate, spending time with his family and taking long walks on Marine Drive. In the midst of this enforced holiday, he reached out to Nandan to find out what was happening with the Aadhaar programme. Nandan replied, 'Why don't you come to Delhi and get started?' Viral landed in Delhi, spent his first day there attending a meeting at the offices of India Post, and within the month found himself working as an unpaid volunteer at UIDAI—all thoughts of MIT and fellowships now forgotten. It would be six months before he got an official title—Manager, Financial Inclusion—and a pay cheque.

Designing the world's largest social security scheme

With his characteristic wry humour, Shankar Maruwada tells us a story from his time in the trenches at UIDAI. 'As part of the marketing team, we made a real effort to reach out to the people and understand whether they would be interested in enrolling for Aadhaar. Naman Pugalia, one of our volunteers, spoke to a villager standing in line to enrol for Aadhaar in the forty-degree Celsius summer heat of north India. The villager said that although he already had three other forms of ID, he was still waiting patiently to get a fourth. When Naman asked him why, he replied, *"Budbak, agar hamare paas teen bhains hai, aur sarkar hamein chauthi bhains de rahi hai, to hum nahin khade honge?"* (Idiot, if I have three buffaloes, and the government is giving me a fourth, won't I stand in line?)'

This 'a-ha' moment, the realization that the Indian people valued an ID as an economic asset, was especially important for Aadhaar. A programme that's meant to deliver unique IDs to 1.2 billion people, and which places no restrictions on what that ID can be used for, must necessarily be outside the realm of the ordinary, and its success largely depends on its acceptance by the people. A great deal of thought went into the design of both the Aadhaar scheme itself and the organizational structure tasked with its execution, in the process creating a new template for technology-enabled projects within government.

What are some of Aadhaar's unusual attributes? To begin with, it is not mandatory to possess an Aadhaar number—residents can choose to enrol voluntarily, and no government service or benefit is allowed to mandate the use of Aadhaar for identity verification. The second is its openness, exemplified by the fact that Aadhaar is designed as a platform providing a single service—identity verification—that can be easily plugged into any application requiring such a service. Today, Aadhaar is used to verify identity in a host of government schemes and services. The Mahatma Gandhi National Rural Employment Guarantee Act now uses Aadhaar numbers to make payments; recipients can withdraw money through Aadhaar-linked microATMs. The subsidy for Liquefied Petroleum Gas (LPG) is administered by linking a consumer's information with their Aadhaar numbers. Banks have established e-KYC (electronic-Know Your Customer) processes using Aadhaar to open new bank accounts, and the government itself uses Aadhaar to track the attendance of employees.

Keeping it simple

A few months after his appointment as chairman, Nandan met K.V. Kamath, then the chairman of ICICI bank, to deliver a presentation about Aadhaar and its uses. At the end of his talk, an amazed Kamath declared that the entire scheme boasted of a 'diabolical simplicity'. Part of the reason for this simplicity was purely practical—if you have to collect 1.2 billion data sets that will be compared against each other every time a resident uses their Aadhaar number, it's best to collect the least possible amount of information.

Pragmatism also dictates that the path to success is easier if you provide a 'thin' solution—one that does not infringe on turf that other government agencies lay claim to. Aadhaar provides a single, clearly defined piece of information—a person's identity—and nothing more. This minimalistic approach tipped the scales in favour of collaboration over competition. Equally important, the agencies that chose to use Aadhaar were given the freedom to decide how they wanted to deploy it, whether it was for opening bank accounts or distributing pensions. Providing this level of freedom also helped to dramatically reduce the friction between the stakeholders in the system.

Aadhaar's simplicity was not restricted to its use; it was clearly manifested in the enrolment process as well. We wanted to provide residents with the kind of smooth and seamless experience they have come to expect from the private sector. With this goal in mind, people were given the freedom to enrol in any manner and at any location that was most convenient (we explain this enrolment model in greater detail in the following chapter). We also granted ourselves the same freedom of choice when it came to building the technology platform for implementing Aadhaar.

Another important design principle we adopted was that of asynchronicity—every part of the Aadhaar ecosystem was designed to function independently. For example, states could choose to begin Aadhaar enrolments at their own pace. Registrars could join the system and enrolment agencies could scale up their activities as per their convenience. This asynchronicity extended to Aadhaar's technical platform as well. Whether internal or external, there was no dependence on any one critical step which, if improperly executed, could bring the whole process to a crashing halt.

Any citizen-centric design in India today must take into account the shifting demographics of our country. With an increasingly mobile population, we need services that are no longer geographically tethered to one location and which are easily portable. Many government services are perceived to be citizen-unfriendly because the fear of fraud has made entry into the system inconvenient at best and impossible at worst. Aadhaar was meant to reach precisely those people who had traditionally been excluded from government systems due to lack of documentation; we harnessed the power of technology to lower entry barriers, promoting inclusivity, while setting up fraud detection systems at the back end to safeguard data integrity and weed out malpractice.

Building partnerships

For fourteen months, while the UIDAI core team laboured to build the organization and the technology to start generating Aadhaar numbers, Nandan travelled to every state of India on what he calls his 'yatra', meeting the chief minister, the chief secretary or other senior officials to get their personal commitment to the Aadhaar project, operating on the principle that an entrepreneur always goes to his customers to drum up business, not the other way around.[7] This went to the extent that if Nandan had to meet a chief minister in Delhi, 'I would make it a point to meet him in his Bhavan and not make him come to my office. By doing this, I was signalling to them that I care about you and I am coming to you because I want your help in doing something nationally important.' This small touch was enough to melt any resistance and get their support. Every chief minister saw a different value in the Aadhaar programme, depending on the specific needs of his or her state. For example, Nitish Kumar, then the chief minister of Bihar, appreciated the fact that Aadhaar was designed to address the needs of a highly mobile population, a major concern in a state from which people routinely migrate for education or employment.

The same philosophy came into play to get other major players on board. In the financial sector, Nandan had started establishing contact with such organizations as the Reserve Bank of India (RBI), the State Bank of India (SBI), the Life Insurance Corporation of India (LIC) and others even before his official appointment as UIDAI chairman. He and members of his team met representatives of these organizations and many others—the Indian Banks Association, the Securities and Exchange Board of India (SEBI) and India Post, to name a few. Simultaneously, he was also reaching out to all the central ministries, the railways and the defence establishment. Nandan reminisces, 'A surprise for me was how stiff and hierarchical the system was. Information reached you only through "official" channels, levels of hierarchy were strictly observed and you were never supposed to go to an office of someone junior, or be on the line when he calls. I decided to break out of all that and reached out to everyone, from Cabinet ministers to their private secretaries.'

Other stopping points on his evangelization tour were commercial organizations like telecom and oil companies, multilateral agencies like the World Bank and the Asian Development Bank, bilateral agencies like the Department for International Development, and various media outlets. This helped to create a powerful coalition that had a stake in the success of Aadhaar, and the value of these efforts was seen when Aadhaar enrolments finally began. State governments across the political spectrum signed up for their residents to be enrolled for Aadhaar, and a huge ecosystem with multiple public and private partners sprang into existence to leverage the power of the Aadhaar platform.

Ranjana Sonawane's Aadhaar enrolment in September 2009 signalled an entry into the phase where operations needed to be scaled up. After flagging off the first enrolment, the UIDAI raced to reach a run rate of one million enrolments a day. In order to do so, MoUs were signed with registrars, enrolment agencies were brought on board, operators were trained, a biometric device ecosystem was created, enrolment and de-duplication software was continuously upgraded and fine-tuned to function at scale, servers were procured, letters were printed and dispatched, and a multilingual call centre was set up to handle queries and grievances. It took a significant amount of time and effort right from the first enrolment to scale each of these processes to achieve the target of generating one million Aadhaar numbers a day, but when Nandan tendered his resignation as chairman of the UIDAI on 13 March 2014, he could do so with the knowledge that the UIDAI had succeeded in its goal of delivering over 600 million Aadhaar numbers in less than five years of its existence. In fact, over 900 million residents have been registered for Aadhaar in the five years since Ranjana Sonawane received her Aadhaar.

Opposing voices

The UIDAI faced a number of challenges in its early days, as the accompanying diagram explains. Perhaps the biggest was from the home ministry, responsible for creating the National Population Register (NPR), a scheme that proposed to collect biometrics and issue a smart card to every citizen of India.[8] Aadhaar, on the other hand, was designed purely as a number for identity, carrying no information on the holder's citizenship status. The UIDAI had also got off to a faster start than the home ministry; by the time Aadhaar enrolments had begun, the Registrar General of India (RGI) had only carried out a few pilot studies for the NPR. Even so, the idea that two separate government agencies would collect the biometric details of all Indian residents for two separate schemes was seen as a waste of government money by the Department of Expenditure in the Ministry of Finance.

Given the overlap between the two projects, the home ministry wanted the UIDAI to stop enrolling residents, and instead act as a back-end organization, collecting and sharing all data with the NPR. The RGI and the home ministry felt that they were the agencies responsible for running any sort of identity scheme, and that the UIDAI was an unwelcome intruder into their turf, thus rousing their protectionist instincts. Eventually, a consensus was reached and the RGI agreed to become a registrar of UIDAI, in effect enrolling residents for the Aadhaar number. The government directed the UIDAI to carry out enrolment in half the states of India and the RGI in the other half. The software of the UIDAI and the NPR were also made compatible so that the data collected for Aadhaar could also be used by the NPR. This détente proved beneficial to both the projects, and nearly a fourth of all those who have received an Aadhaar number have been enrolled by the RGI.[9] Despite initial

THE AADHAAR OPPOSITION WHEEL

Opposing voices against Aadhaar

LEFT-WING

'Biometric data collection compromises privacy and security'

Aadhaar collects minimal demographic and biometric data. Data is stored securely.

'The state will abdicate its role in social welfare'

Direct cash transfers via Aadhaar help to clean up existing social welfare systems.

RIGHT-WING

'Enrolling for Aadhaar grants de facto citizenship'

Aadhaar only acts as a proof of identity and address, not citizenship. The government decides who a citizen is, independent of Aadhaar status.

COMMERCIAL VENDORS

'Only smart cards can guarantee security'

Solutions using smart cards end up being too complex. Aadhaar is simple and scalable because it is not tethered to a physical form while still being secure.

VESTED INTERESTS

'Direct cash transfer for LPG subsidy is unworkable'

Direct transfers eliminate diversion and black-marketeering. Detection and removal of duplicate connections alone could save 1.93 billion rupees a year.

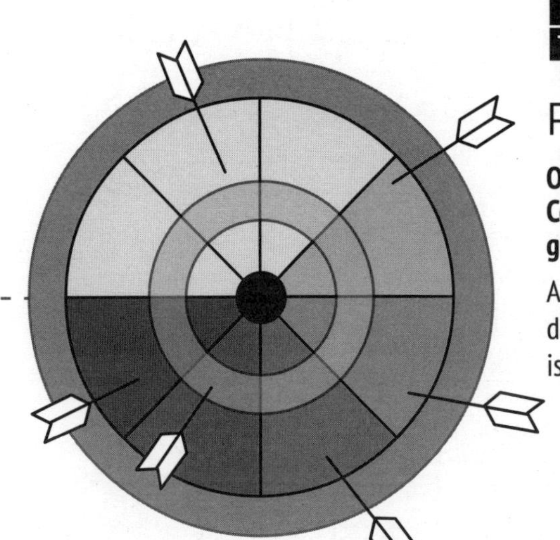

PARLIAMENT

Opposition from the Standing Committee on Finance on legal grounds

Aadhaar is compliant with existing data protection standards. Aadhaar is not a proof of citizenship.

LEGAL OPPOSITION

'Benefits will be denied to people without Aadhaar'

Aadhaar has never been mandatory to receive benefits.

'Aadhaar data can be shared with other government agencies'

An individual's Aadhaar data cannot be shared with anyone without their explicit consent.

BUREAUCRATIC

Turf wars with the National Population Register

After initial clashes, the NPR and Aadhaar worked together.

'Iris data collection is expensive and unnecessary'

Iris data capture adds 4.4 rupees to the cost of a single enrolment. Without iris data, fingerprints alone cannot provide 1.2 billion unique identities.

opposition, the then home minister P. Chidambaram was later quoted as saying that Aadhaar was a tool of empowerment for 'those at the bottom of the pyramid, the poor, the migrant workers, the homeless and the oppressed'.[10]

The next speed bump centred around the UIDAI's decision to use biometric authentication, specifically the use of iris scans. The Planning Commission formally raised the issue of whether the additional expenditure of capturing the iris data was justified.[11] Expert members of the Biometric Standards Committee constituted by the UIDAI had suggested that fingerprint data alone would not be robust enough on which to base the promise of 1.2 billion unique identities. Fingerprints can wear out due to physical labour and are unstable in the case of children younger than sixteen. On the other hand, iris patterns are nearly fully developed at the time of birth and remain constant throughout an individual's life. Despite what we may have seen in the movies, it is very difficult to duplicate someone's iris, and multiple international security systems (such as border control and immigration) routinely capture iris data for security purposes.

Given all these benefits, a cost analysis was performed by the UIDAI. Capturing and de-duplicating iris data was found to increase the cost of each enrolment by around 7 cents (Rs 4.40). Of this additional sum, the capture device cost 4 cents (Rs 2.90) while the additional labour as well as the iris de-duplication software cost 75 paise each. These numbers were projected to decrease as enrolments went up. The overall increase in the project cost was estimated at $77M (Rs 5 billion), an expenditure justified by the massive improvements in inclusiveness and data security that iris data can bring about.[12]

It wasn't just the government that raised objections to Aadhaar. Concerns about national security led some people to protest on the grounds that any resident of India could enrol for Aadhaar, even non-citizens; they were apprehensive that illegal immigrants could use their Aadhaar number to bilk the government of funds by claiming social security benefits. These fears are not grounded in reality. As we've pointed out before, Aadhaar is purely an identity platform, and provides no information other than verifying who someone is. It offers no information on the citizenship status of the Aadhaar holder, and merely furnishing your Aadhaar number is not sufficient to get any welfare benefits; you still have to prove to the state that you are below the poverty line if you want subsidized foodgrains, or that you are above eighteen and live in a rural area if you want to enrol in the MGNREGA scheme, or that you are a child enrolled in a government school if you want a free mid-day meal, and so on.

Although the Aadhaar number was never meant to be mandatory for any government service, some felt that no government schemes should be linked to Aadhaar until all 1.2 billion residents of India had one. They were concerned that prematurely making an Aadhaar number-based service available would leave the unenrolled out in the cold. The UIDAI has always advised government agencies to rely on alternative methods of identity verification until Aadhaar achieves sufficient coverage. Unfortunately, some agencies jumped the gun in declaring the Aadhaar number to be mandatory for availing of certain benefits. This was in opposition to the UIDAI's stand on the matter, and the Supreme Court had to step in and reverse any such declarations.

Another point of view held by some NGOs was that Aadhaar-linked direct benefit transfers could be used as an excuse by the state to abdicate its responsibility towards social welfare. Instead of building better schools and primary healthcare clinics, for example, activists feared that the government would restrict itself to disbursing cash benefits using Aadhaar. However, using Aadhaar to accurately identify those individuals deserving of social welfare benefits has been acknowledged by the present government as an essential step towards welfare reforms, and has always been a key goal of the programme.

Citizen groups concerned with data privacy and security objected to the idea of creating a database containing the demographic and biometric data of all Indian residents, fearing the emergence of a surveillance state in which biometric information could be used as a targeting mechanism; at a time when whistle-blowers like Edward Snowden have revealed the extent to which governments can snoop into the private lives of unsuspecting citizens, such a database might prove an irresistible temptation to a government wanting to keep tabs on its people, violating their right to privacy in the process. Given that India still doesn't have a single, well-defined law that regulates data privacy, such concerns were valid. The UIDAI was well aware of these issues, and hence designed all of Aadhaar's data collection, storage and retrieval processes with great emphasis on security. These design decisions have been outlined in great detail in many of the technical documents put out by the UIDAI.

Once the UIDAI collects an individual's data, it is stored securely in government-owned data centres in Bangalore and Noida. Unless an individual provides explicit consent, their data cannot be shared with anyone, not even another government department. If any government or private agency wishes to make use of Aadhaar for identity verification or e-KYC processes, they can only do so with the consent of the Aadhaar holder. In such cases, the information supplied by the UIDAI to the agency is exactly the same as the data found on any other government-issued photo identity card.

This stance on data privacy was tested in 2013, when a Goa court ordered that the Central Bureau of Investigation (CBI) be provided access to the UIDAI's database. A Goan schoolgirl had been raped, the police were unable to solve the case, and the CBI hoped that the database could be used to compare fingerprint data and identify possible suspects. Although well intentioned, such a directive was in clear violation of the UIDAI's policies on data privacy and individual consent. The UIDAI ended up petitioning the Supreme Court on the matter, who ruled that no such information could be provided to the CBI unless the suspect himself had authorized it, in effect barring law enforcement and other government agencies from obtaining an individual's biometric data without their approval.[13]

Many of these issues were resurrected when the National Identification Authority of India (NIDAI) Bill was proposed. This was the piece of legislation that would grant the UIDAI statutory approval and parliamentary recognition. The Standing Committee on Finance presented a parliamentary report regarding the NIDAI Bill which rehashed the older arguments against Aadhaar.[14] One contentious issue was the distribution of power between the executive and the legislature when it came to overseeing a project of this magnitude. The hazard of duplicate identity schemes run by different governments also found a mention. As we have described, a consensus was reached between the UIDAI and the home ministry regarding the distribution of work for the Aadhaar and NPR schemes. The UIDAI also reiterated that Aadhaar is meant to function as an open platform, compatible with any scheme that requires ID verification. Some of the technical issues regarding the accuracy of biometrics as discussed in the report were factually incorrect. The UIDAI has issued several detailed reports that refute these arguments. As of now, the NIDAI Bill has not been enacted as a law by Parliament. However, the new government at the Centre has thrown its weight behind the Aadhaar scheme, extending enrolments to an additional four states—Uttar Pradesh, Bihar, Chhattisgarh and Uttarakhand—and setting the UIDAI a target of a billion Aadhaar enrolments to be completed in 2015.[15]

A blueprint for innovation in government

Whether it's giving Aadhaar numbers to two-thirds of the country in five years, or taking on any other project that seems audacious at best and impossible at worst, some principles always hold true.

Set a goal that is reachable, but which requires a lot of hard work and effort to get there. Have a clear-cut vision of what you are trying to achieve, a vision that should be inspiring enough and powerful enough to attract the best minds in the business. Try and build as diverse a team as you can, and don't let your thinking get bogged down by whether they're bureaucrats, tech geeks, finance wizards or academicians, as long as they have something of value to contribute.

Many of the lessons we learnt in these five years were probably unique to the environment in which Aadhaar came to be, but are generalizable beyond Aadhaar. In the din of democracy, every ministry, committee, agency and member of Parliament has a viewpoint on a given programme, and all these viewpoints must be considered and given due attention. We must be sure that the needs of all groups are being met and nobody is being inadvertently excluded. This can only be achieved if we can build a coalition for change.

Despite the opposition that the Aadhaar programme faced from some quarters, the UIDAI worked hard to build a strong, positive coalition that understood and supported the goal of granting an identity to every Indian. This coalition had many faces, each with their own reasons for participating. There were state governments, who recognized that Aadhaar and its various applications could benefit both the administration and the people under its jurisdiction. For every smart-card supplier who was unhappy, there were biometric vendors who supported Aadhaar's biometric efforts. While some NGOs rose in protest, there were others that endorsed Aadhaar as a solution for the poor and marginalized to gain an identity. And finally, there were the people of India, who voted with their feet and walked in large numbers to their local enrolment centre to voluntarily sign up for an Aadhaar number—the silent millions who finally spoke and claimed their unassailable right to possess an official identity as a resident of the country.

In any private sector company, there are a limited number of people to whom you are answerable. In the government, however, every project has many masters, all of whom can, and often do, question the manner in which things are being done. The former commissioner of the US Internal Revenue Service (IRS), Charles O. Rossotti, includes in his brilliant book *Many Unhappy Returns* a sketch that Senator Bob Kerrey drew for him on a scrap of paper during his confirmation hearing, listing out all the government officials and departments to whom Rossotti, as the IRS commissioner, would be directly answerable. There are sixteen different agencies on that list. Not surprising that Senator Kerrey signed his artwork, 'Good luck in the bullseye'. It was no different for the UIDAI, which was answerable to multiple departments and agencies within the government as well as to the public at large. We had to learn how to handle feedback from so many different sources, and assimilate all these different opinions.

The UIDAI began operations with a minimal staff housed in temporary offices in New Delhi and Bangalore, operating on a shoestring budget (we discuss this in detail in the next chapter). We could have waited for years until formal offices were built, a full complement of officers was appointed and an ironclad legislative bill was passed before we started work on enrolling residents. Had we done so, it is entirely possible that today not a single Indian would have received his or her Aadhaar number as yet. Forging ahead was a calculated risk, but one that bore rich rewards. The many innovative ways in which Aadhaar is being deployed would have otherwise remained a mirage confined to the pages of government reports.

Let's return to where we started, to Ranjana Sonawane of Tembhli. Did her life take a turn for the better after she received her Aadhaar number? When reporters visited her in 2014, not much had changed. She was still struggling to make ends meet, concerned about providing a good education to her three sons. She asked, pointedly, 'How will I eke out a living with a card?' Her question highlights the fact that Aadhaar is not a panacea for all that ails the many Ranjanas across the nation. If our country is a jigsaw, then Aadhaar is only one piece of that puzzle. Ultimately, Ranjana's story tells us that it will take multiple government systems working in tandem to build a picture of a truly inclusive India that fulfils the aspirations of all its citizens.[16]

2

AADHAAR: BEHIND THE SCENES

LIKE MANY OTHER bureaucrats, Ram Sewak Sharma is possessed of a polite manner and speaks beautifully chaste Hindi. However, behind this courtly veneer lies an extremely sharp, pragmatic and decisive mind, fairly unusual in government circles. Also unusual is his keen understanding of technology, honed during his days in the Jharkhand state government's information technology department. A hobbyist programmer since college, he'd even taken a mid-career break to get his master's in computer science. On 1 July 2009, he met Nandan at the Maurya Sheraton in New Delhi to discuss what he had been told was a 'project of national importance'. As he tells us, 'Nandan informed me that he was going to join the UIDAI as chairman and was looking for somebody to work as CEO. We discussed the project in general terms, but from a bureaucrat's perspective it was clear that this was a brand new start-up and would have to face huge challenges in setting up the initial infrastructure and organization. When Nandan asked me if I would like to do it, I said yes, because if he could take the risk, so could I.'

Six months earlier, the UIDAI had officially come into existence, and the government had reached out to Nandan with the opportunity of leading this programme. While exciting, it was also an extremely challenging project, completely unlike anything the government had worked on so far. And so, Nandan reached out to several colleagues and experts in various fields to gather their opinions. Some of these colleagues included Srikanth Nadhamuni, a co-founder of eGovernments Foundation, a non-profit organization that helps governments improve service delivery via technology; T. Koshy, then an executive director at the National Securities Depository Ltd, who was familiar with the use of technology in government; and Sriram Raghavan, the founder of Comat Technologies, a firm that had worked extensively at the ground level with ration-card issuance and duplicate elimination in Karnataka. These early meetings familiarized Nandan with the kinds of challenges and complexities to expect, and were key in forming the early architecture of Aadhaar.

Next, Nandan sought a meeting with Montek Singh Ahluwalia, deputy chairman of the Planning Commission, the government department that was to be the official home of the unique ID scheme for the next five years. All these discussions convinced Nandan that building a unique ID platform was exactly the sort of worthwhile challenge he could sink his teeth into. On 15 June 2009, he met the prime minister Manmohan Singh, and officially expressed his interest in heading the unique identity project. Ten days later, he was appointed chairman of the UIDAI at a Cabinet meeting. By 9 July, he wrapped up his responsibilities at Infosys; less than six weeks after his first meeting with the prime minister, Nandan took charge of the organization that would implement the 'biggest social project on the planet'.[1]

An aggregation of talent

On 12 August 2009, the prime minister's advisory council held its first meeting with the objective of establishing the UIDAI's initial benchmark—the number of people who would receive an Aadhaar number within the next five years. In order to come up with an estimate before this meeting, the UIDAI organized a conference with experts from various fields. The economists Abhijit Banerjee and Esther Duflo, B.Venugopal, an executive director at the LIC, Shankar Maruwada, a marketing and analytics expert, government officials such as Ram Sewak Sharma, professors from Ivy League colleges, representatives from NGOs and many others weighed in. The initial estimate that emerged predicted that 400 million people could be enrolled in five years. Nandan felt that this was too conservative a goal. As he says, 'This is one-third of our population and appeared too small given that we had five years, so I said let us target 600 million. When in the meeting I committed 600 million by March 2014, I had no clue how to get there. It's a happy situation that by the end, we comfortably surpassed it.'

Setting such an ambitious goal was a deliberate choice; during his time in the private sector, Nandan had come to realize the value of stretch goals that could push people out of their comfort zones and energize them into delivering the impossible. How could this principle be put into practice inside the Byzantine byways of the Indian government? The original proposal envisioned the UIDAI being set up along the traditional lines of a government agency, with every state having a joint secretary heading the operations and managing a large field staff, all potentially adding up to thousands of employees. Instead, Nandan decided to take up an approach he describes as a 'crazy idea'—to run Aadhaar as if it were a start-up inside government, following in the footsteps of trailblazing organizations like the Atomic Energy Commission and the Indian Space Research Organization. The UIDAI adopted a lean, open organizational structure—key planning and strategy decisions were taken by a core team that numbered only 300 at most. The actual execution of the project became the responsibility of outside vendors, who were selected to partner with the UIDAI through standard government processes.

The core team needed to have people who were comfortable with risk, who had expertise in multiple fields, and who had both the energy and the enthusiasm to chase down a formidable target. It turns out that the Indian government had many such individuals within its ranks. Following in the pioneering footsteps of Ram Sewak Sharma was K. Ganga, Deputy Director General (Finance), whose crisp saris became a personal trademark. She came from the Indian Audit and Accounts Services and had been at the forefront of pioneering IT-based initiatives as far back as 1989. She also served as the financial advisor to the President of India. She recollects, 'When a common friend asked me whether I would like to join Nandan, the IT czar of India, as his chief financial officer for the most ambitious and transformational project that was ever undertaken by the Government of India, I did not even bat an eyelid before I said yes.'

Nandan's private secretary, M.S. Srikar, a gold medallist from the National Law School of India, was originally from the Karnataka cadre of the Indian Administrative Service. 'Having worked in the government for over little over a decade in 2009, the opportunity to work in a new, unique experiment which could redefine the way government functions was an exciting prospect. I was undergoing my mandatory mid-career training in the IAS Academy in Mussoorie when I was informed that I was proposed as the private secretary to the chairman of the UIDAI, Mr Nilekani. It was followed up with an interview in Delhi at short notice before I was picked. It was rare to be interviewed in government for an assignment, but Nandan put me at ease and sought my views on a number of issues,' Srikar recounts. In order to help with the complex technology procurement process within government, B.B. Nanavati joined the UIDAI from the Indian Revenue Service, bringing with him knowledge of years of IT procurement.

Ashok Pal Singh, who served as deputy director general (DDG) in charge of financial inclusion and also logistics, came into the UIDAI from India Post. Soft-spoken and businesslike, Viral describes him as a 'rare strategic thinker in a tactical environment'. The idea of a unique identity for every resident of India was something that he had devoted much thought to, going so far as to come up with a plan for post offices to provide ID cards and bank accounts to people via a smart card-based system. When he heard of Nandan's appointment, he remembers thinking, 'Here is a person who will buy into this idea unlike regular people in the system.' He sent Nandan an unsolicited email—'a shot in the dark', he says—and got a reply within fourteen minutes, inviting him to come on board.

Rajesh Bansal, Assistant Director General (Financial Inclusion), was in the US completing a master's in public policy at Duke University; his thesis dealt with the need for India to set up a national ID system to improve governance and service delivery. He tells us, 'I was taking an evening walk when my wife called me from India and told me, you're going to fall off your chair. You will not believe this but the government has just announced the scheme you've been writing about in your thesis. Naturally, as soon as I got back, I was terribly keen to contribute in some way. I shared some of my ideas with Ram Sewak Sharma and he asked me to join the UIDAI.'

The UIDAI's roster was dazzling in its diversity. Government officials were deputed from the Indian Administrative Service, the railways, the postal service, the telecom service, the Reserve Bank of India, the defence accounts service, the civil accounts service and the revenue service, among others. Experts from the private sector came on board as well, with experience in technology, marketing, public relations, outreach to civil society, law, research and academia, and other related fields. Many such individuals were hired as consultants through the National Institute of Smart Governance (NISG). Others, especially members of the technology team, were initially hired as volunteers or took sabbaticals from their organizations, including such technology behemoths as Hewlett Packard, Oracle and Intel. In fact, Nandan wrote a letter to NASSCOM (National Association of Software and Service Companies), asking them to lend their employees to the Aadhaar project. The response was extremely positive, with many companies contributing talent to the Aadhaar technology team at their own cost. Students from some of the world's best educational institutions were selected as interns. Irrespective of their designation, everyone who joined the UIDAI had to undergo an interview and sign an agreement. Everyone had a clearly defined role and responsibility within the organizational structure.

Many of those who joined gave up comfortable positions and perks to be part of this project. Ashok Pal Singh summed up the situation perfectly when he said, 'In terms of perks, privileges and pay, this place has nothing more to offer than any job in government. In many respects, it is worse off. And yet, everyone is here because they want to be.' Without this diversity and this level of commitment, the Aadhaar project may never have got off the ground.

Turf wars between government ministries and a lack of coordination mean that projects developed in one ministry may not be adopted by others. To avoid these sorts of squabbles, the UIDAI was housed in the Planning Commission (the present-day Niti Aayog), a neutral body. This way, all ministries could sign up for the Aadhaar platform without stepping on any toes.

The government births a start-up

In theory, the birth and evolution of the UIDAI sounds quite orderly and bureaucratic in nature—deciding how to execute a massive government project, defining some key operational principles, setting up committees to issue recommendations, and roping in consulting experts for their opinions. But all of this orderly thinking happened not in spacious government offices, staffed with a full complement of personal assistants, but in the chaotic, frenetic environment familiar to anyone who's ever set foot in a technology start-up.

The UIDAI started functioning with a bare-bones budget of $15M (Rs 1 billion) for the first year (which was never completely utilized), a skeleton staff, and very little office space. And yes, 1 billion rupees in the Government of India is considered a small project. The senior bureaucrats who held the posts of director general (DG) and deputy director general would have been entitled elsewhere to comfortable cabins of their own; here, they were squashed, cheek-by-jowl, with everyone else in a miniscule room in the erstwhile Planning Commission's Yojana Bhavan office in New Delhi. Eventually, the UIDAI headquarters moved to the Jeevan Bharti building in Connaught Place, a location formerly occupied by Air India. Even then, the team did not wait for the offices to be renovated before they moved in. They worked in one half of the office, ignoring the noise and dust around them as the other half was being refurbished.

Simultaneously, the technology team began to take shape in Bengaluru. Passionate about wanting to make a change, entrepreneurs from Silicon Valley and employees of blue-chip technology companies alike signed up to participate. Leaving their families behind, some of the early team members moved to Bengaluru, camping in an apartment on the Outer Ring Road, a stone's throw away from the house of Srikanth Nadhamuni, the UIDAI's head of technology. The living room of that apartment became the Bengaluru tech team's office for several months, with many of the key design decisions being outlined, in Srikanth's words, 'on two whiteboards and at two tables'. Raj Mashruwala, soft-spoken and a meticulous planner, was one of the project's earliest volunteers. He tells us, 'I had signed up to be part of the UIDAI project around the end of July and flew back to the US to tell my family I was moving to India. I came back two weeks later, took a taxi from the airport and met Srikanth at an appliance store on Outer Ring Road to buy the necessities for the apartment. The next day the office was operational.' The UIDAI's chief architect, Dr Pramod Varma, also joined the team in the very early days. A member of the royal family of the former kingdom of Travancore, his boundless enthusiasm and boyish mop of hair give him the air of a college student. Having worked with Nandan at Infosys almost a decade earlier and subsequently in start-ups, he recalls, 'The moment I read about Nandan having been appointed to build a national ID platform, I knew that I wanted to be part of this project, and immediately reached out to him.'

Vendors who visited the team expected the trappings of a government office, and were utterly perplexed to find that they had apparently wandered into a typical bachelor pad. Raj recalls, 'We gave the guys from Sun Microsystems our address and they called us up once they reached to ask if this really was the correct place or if they had got lost. Another time, we met a team from one of the biometrics vendors at the apartment. Later, they told us they could not believe that this was an official Government of India organization—they actually wondered if it was an elaborate sting operation set up by a competitor to trick them into revealing their technology.' Perhaps to the relief of the technology vendors, the team later moved into more formal offices in one of Bengaluru's mushrooming tech parks on the nearby Sarjapur Ring Road.

The situation was ripe for culture clashes between those who had spent years in the hierarchical, highly formal bureaucratic structure of the government, and those from the much more informal private sector background, where shorts were considered office wear and it was not just acceptable but expected that you called the CEO by his first name. One world changes at glacial speeds, while the other metamorphoses overnight. On the first day he joined UIDAI, Viral attended a meeting at the Department of Posts. 'Here I was, having just returned to India after a decade in the US, sitting in the conference room surrounded by senior bureaucrats from the Indian postal service. Then the chairperson of India Post walked in, and to my absolute amazement, everyone in the room stood up!' he recalls. He had to learn the rules as he went along; you couldn't just barge into an office in Delhi, but had to enter according to order of seniority. In a meeting, you couldn't just speak up whenever you had a question; your rank determined when it was appropriate for you to talk.

Viral himself is fairly outspoken, bluntly direct where bureaucrats tend to be oblique and hierarchical. Only half in jest, Ram Sewak Sharma bestowed upon him the title of 'free radical', both a description as well as a clever chemistry pun. This nickname stuck and became widely adopted, and today Viral laughs about it. 'I was allowed to do a lot of things because people said, "Oh, this guy's a free radical, he's going to come up with something different!"'

The bureaucrats were equally perplexed by these brash private sector fellows, who didn't seem to understand the concept of expense records, public audits or paper files. Each group had their own impenetrable jargon—talk of government financial rules, committees, questions from members of Parliament and due process, was countered with the language of cloud storage, servers, bandwidth and data encryption. Those in the private sector dashed off quick emails to communicate with team members, largely unconcerned with building institutional memory, whereas the holy grail for communication and record-keeping in the government continued to be the quaintly anachronistic file containing green sheets and ties on either side, wending its ponderous way through the bureaucratic hierarchy. One set of people wanted to do things first and worry about them later, whereas the other set was always mindful that they would be held accountable for every step in the decision-making process.

Despite the potential for a comedy of misunderstandings, the entire organization overcame some initial hiccups and settled down to function smoothly.

Both sides realized that they had to learn each other's vocabulary, and more importantly, earn each other's trust. Given the magnitude of the challenge at hand, they could not afford to function in silos, and adopted a collaborative model of work where shared expertise helped to drive innovative decision-making. Viral recollects Nandan telling him at one point, 'When you set yourself a goal that large, everything appears small in comparison.'

The organization continued to grow, and eight DDGs were appointed, reporting to the DG. Reporting to each DDG were two to four assistant director generals (ADGs). Eight regional offices were created around the country to work closely with the state governments for Aadhaar enrolment and customizing the rollout to the local requirements. These offices were set up in Mumbai, Bangalore, Hyderabad, Ranchi, Guwahati, Chandigarh, Lucknow and Delhi. At its peak, the entire organization consisted of about 200 government officials, and another 100 or so consultants, volunteers, those on sabbaticals, and interns.

In the meanwhile, the technology team in Bengaluru was working furiously in the Ring Road apartment, trying to meet the goal of issuing the first Aadhaar number by 2010. While the core of the technology team was in place and design decisions were being finalized, the programmers who would build the entire system had yet to be brought on board. The sense of urgency to achieve the one-year target was signalled by the fact that the DG, Ram Sewak Sharma, didn't waste time waiting for programmers to be hired to work on the enrolment software; he rolled up his sleeves and wrote the entire thing himself in a matter of days, working after office hours and on weekends.

In 2013, roughly four years after the birth of the UIDAI, Ram Sewak Sharma stepped down from his post as DG. This marked the end of the start-up phase and the next DG, Dr. Vijay Madan, worked to scale up Aadhaar across the country. Enrolments were increased in states that were lagging behind, and the UIDAI worked with a number of ministries and state governments to build Aadhaar applications.

A new database for a new India

Amelia, in her early twenties, had one simple desire—she wanted to buy a two-wheeler. Money was not the problem, since she had scrimped and saved for the past three years. The problem was identity. Sold into prostitution before being rescued, she lived in a protection home in Chennai without identity documents of any kind. Fellow resident Priya wanted to work at a hotel in the Middle East, but couldn't get a passport since she too had no identity documents.[2]

Both of them were among the nearly 500 women and children, part of a special camp organized by the UIDAI, who received the first-ever identity document they had possessed—an Aadhaar number. In Jharkhand, women who had received Aadhaar cards felt more empowered; they could now collect government benefits on their own, without relying on their menfolk. Their status in the community went up, and some women used this boost to tackle other social issues in their villages, like getting their husbands to stop drinking.[3]

How was it that all these people qualified for an Aadhaar number, despite being unable to obtain any other form of identification? The answer is that Aadhaar was deliberately designed to be as inclusive as possible, especially targeting those who had fallen through the cracks of all other government identity schemes. Long before a single piece of data was collected from any Indian resident, the UIDAI put in hours at the drawing board, planning exactly what type of data it would need to collect to succeed in its mission of inclusiveness.

The first question was as simple as asking whether the UIDAI needed to enrol any residents at all. Was there an existing database they could use for the job? The answer was, not quite. While some official databases did exist—voter rolls, the list of people who fall below the official poverty line and qualify for benefits—they had several major drawbacks. One, the data they contained was not sufficient to ensure a truly unique identity for every Indian resident, and was often beset with errors. Two, they did not cover the entire population of the country. And three, each of these data sets had their own structure and format; integrating them all would be a logistical nightmare. All in all, it was clear that the UIDAI would be better off building its own de novo database, with all data being collected in one shot.

The next decision to be made was as to the kind of data to be collected. It was clear that there would be two types: demographic and biometric. To handle the demographic side of things, the Demographic Data Standards and Verification Procedure Committee was set up, chaired by N. Vittal, former central vigilance commissioner, which subsequently came up with an exhaustive set of guidelines for demographic data collection.

Given that every field collected would have to be repeated 1.2 billion times, the effort was to keep things as simple as possible. But nature abhors a vacuum, and various government departments jumped in to try and fill the perceived gaps in the system. The health ministry suggested that the UIDAI collect the blood group data of each individual; other ministries wanted information on caste, disability status and other attributes. Ram Sewak Sharma reminisces, 'A common mistake that government applications make is to collect data which they are not likely to use at all, creating long and complicated application forms—the pension form for government servants is like a booklet, with some absolutely irrelevant information like religion and ethnicity. This has happened because the designers of these forms have no idea why they want the extra information, but feel that since it might be required later for some purpose, it's better to collect it now. When we were designing the input form for UIDAI, there were many suggestions from other stakeholders, largely government departments, to collect extra information. The department of minority affairs wanted religion to be collected, and the rural development department wanted income-related information to be collected. The UIDAI adopted a policy that it would collect only minimal information which was necessary and sufficient to establish the identity of a person—

nothing more and nothing less. We just wanted to establish that X is indeed X. We also needed his address to communicate his Aadhaar number to him.'

Eventually, the only fields made mandatory were a person's name, address, date of birth and gender. These mandatory fields are sufficient for the purposes of verifying one's identity and address. The committee report also specified exactly how all this information was to be collected. As a result, data verification and capture have been converted into a standardized process, easily implemented by any one of the many partners in the Aadhaar ecosystem.

Each of these fields was thoroughly discussed and debated before inclusion in the final list. The goal was to make the demographic data as inclusive as possible so that nobody was left out of the database because they didn't fit into a standard category. For example, the Aadhaar number was one of the first government-issued IDs that officially recognized 'transgender' as an option when specifying the gender of the person being enrolled, at a time when most government programmes still forced transgendered individuals to identify themselves as male or female. This option was made available years before the Supreme Court of India officially recognized 'transgender' as a third gender category.

Similarly, it was not mandatory for women to supply their husband's or father's name while applying for an Aadhaar number. Although the field was included in the enrolment process, a woman could choose to leave it blank; she could be issued an Aadhaar number without requiring the data of a male relative to be provided.

Pramod Varma was present at the Demographics Committee meetings where these decisions were taken. He tells us, 'Almost everybody in the committee was initially in favour of including the husband's or father's name while enrolling women. I raised my hand, and the chairman, Mr Vittal, said the young man has something to say, let him speak. I asked, "If this information is made mandatory for all, what about special cases like orphans or homeless people? In practice, anyone who doesn't fit neatly into the mandated categories is often overlooked and eventually excluded from the system." After I finished speaking, I remember Mr Vittal giving a beautiful five-minute speech on why governments become so bureaucratic towards the end user. After some debate, the committee agreed that the best way forward would be to make such details optional rather than mandatory.'

Although the demographic data was restricted to just four fields, capturing even this minimal information turned out to be a Herculean task. The diversity of India manifests itself in the names of its people; there is no standardized naming convention that applies nationwide. People in the south often do not have surnames, but may incorporate the name of their ancestral village and family deity into their names instead. Some people include their father's name as their middle name, others don't. As a result, a person's name may have anywhere from one to five components. Addresses in India aren't standardized either. They often incorporate the name of local landmarks—houses of worship, schools, cinemas, shopping complexes—as points of reference. In India's villages, where everybody knows everybody else, a postal address that says 'behind the banyan tree' or 'deliver to X, son of Y' is often sufficient for a letter to reach its intended recipient.

Eyes, faces and fingers: Identifying Indians

While the deliberations around demographic data collection were ongoing, the same process had begun for biometric data as well, undertaken by a Biometric Committee set up in September 2009.[4] The committee consulted several biometric experts and academics to decide the best type of biometric data to collect and the associated costs. They also closely studied other programmes around the world that used biometric data for various purposes, like the US VISIT programme that collects fingerprints for border security reasons. In parallel, the UIDAI carried out pilot studies to test the validity of fingerprint data, using over 250,000 fingerprint images collected from 25,000 people living across Delhi, Uttar Pradesh, Bihar and Orissa, many of them in rural regions.

Putting together all this information, the best strategy that emerged was to capture a combination of a person's photograph as well as all ten fingerprints and iris scans. The pilot study showed that the agrarian population often has bald fingerprint ridges. If we only used fingerprint data, we would have either excluded the exact section of people most in need of Aadhaar, or our database would have a huge number of errors, since even small errors get magnified with 1.2 billion people. As Ram Sewak Sharma puts it, 'Including iris data in the biometric set was in my opinion one of the most crucial decisions that allowed us to reach 99.99 per cent accuracy when it came to de-duplication. In iris scans, we now have a very powerful authentication factor which especially benefits poor people and in some cases is even better than fingerprint authentication.'

The Aadhaar enrolment pyramid

Once the data collection standards were in place, the next order of business was to decide how exactly a resident could go about enrolling for Aadhaar. The goal was to grant residents the freedom of choice to enrol in the manner most convenient to them, and the Aadhaar enrolment ecosystem that eventually arose was designed to meet this goal. The entire ecosystem can be thought of as a pyramid, with the UIDAI at the top. Next come a class of organizations known as 'registrars', whom the UIDAI appoints. These organizations include all state governments as well as government agencies like the Registrar General of India, India Post, the National Securities Depository Limited, public sector banks and oil companies. The registrars all have one thing in common—they engage directly with residents. Below the registrars in the pyramid are the enrolment agencies, appointed by the registrars after a competitive bidding process in the open market. It is the job of the enrolment agencies to provide the hardware and the operators who will actually carry out the enrolments in the field. Over 100,000 operators have undergone this certification, and at its peak, there were 30,000 simultaneous enrolment stations deployed.

Thanks to this ecosystem, residents could now choose to enrol at a bank, a post office, a government school housing a state-run enrolment camp, or even a ration card shop. Freedom of choice meant that monopolies had no chance to arise; no one could force you to enrol at a particular centre, or pay a bribe to get your number. If residents were unhappy with the quality of service at a particular centre, they could simply go to another one of their choice, or wait for another enrolment agency to set up shop in the same area, as was usually the case.

Recognizing the power of incentives, the UIDAI initially paid the registrars 77 cents (Rs 50) for every successful enrolment. As enrolments scaled up, this figure dropped to 62 cents (Rs 40). Financial incentives also operated in the lower tiers of the ecosystem, where the registrars paid the enrolment agencies anywhere from 38 cents (Rs 25) to 54 cents (Rs 35) per enrolment. The estimated cost of the back-end processes (which did not directly involve the resident)—software development, building a data centre, operations costs, biometric de-duplication, mailing Aadhaar letters—was around 77 cents (Rs 50). Adding up these numbers, the entire enrolment process cost roughly $1.5 (Rs 100) per resident. For a population of 1.2 billion people, the cost is estimated to be $1.8B (Rs 120 billion) over a period of ten years. Compare this to the $46B (Rs 3 trillion) spent by the government on entitlements and subsidies, where a minimum of 10 per cent—$460M (Rs 30 billion) every year—is lost to leakages, fakes, duplicates and ghosts. The numbers speak for themselves. Even so, a detailed cost-benefit analysis of Aadhaar published by the National Institute of Public Finance and Policy found that the project would yield an internal rate of return of nearly 53 per cent to the government.

As with any project of this magnitude and complexity, it was not always smooth sailing. The multiple registrar model faced some operational issues and has been continuously tweaked while leaving the principles of choice and competition intact. Rajesh Bansal recalls, 'When we wrote to all the banks saying that we would like them to be registrars, they were hesitant since they felt this wasn't their business. We had to convince them of the merits of becoming registrars, pointing out that as people enrolled for Aadhaar, the banks could simultaneously open new accounts for them and expand their customer base. Later, banks turned out to be the largest drivers for enrolment in states like Uttar Pradesh, Bihar and Orissa.'

Some state governments wanted to be the only registrars in their state so that they could build a complete database of all their residents. These states wanted the UIDAI to bar all other registrars in their region. Other states, however, did not share this opinion and actively encouraged multiple registrars to achieve full coverage. Ashok Pal Singh recalls, 'Uttar Pradesh was the first state to accept non-state registrars, whereas West Bengal was the first to officially oppose this concept. However, West Bengal's opposition didn't last long, since within a month they issued another notification saying that they were willing to allow non-state registrars to enrol residents of the state.' As a compromise between these two points of view, non-state registrars were eventually allowed to enrol residents only within or around their premises. Initially, public sector banks were not keen on letting private sector banks become registrars. These turf wars were not unexpected, and subsided as registrars came to realize that the power of Aadhaar was in the open platform it provided, rather than in the enrolment process itself. As a result, internecine conflicts came to a halt as enrolments sped up.

Twelve digits for everyone

Here's a deceptively simple question: how many digits should a number have to make it secure, and to ensure you can give out 1.2 billion numbers without any repetition? The answer we came up with was twelve. These twelve digits also need to be completely random so that there's no easy guessing possible. Consider that in the US, you can guess a person's social security number if you know their birthdate and the location from which their number was issued.[5] With identity theft becoming increasingly common, issuing completely random numbers was the only way to combat such issues, so much so that even if a family enrols for Aadhaar together, each member will get a number that is totally different from the others.

We said twelve, but Aadhaar is actually eleven random digits, followed by a twelfth digit whose purpose is to detect data-entry errors. This numbering system is similar to the one used for credit cards, where a formula called the Verhoeff Scheme is used to calculate the last digit. The twelve-digit format allows for enough permutations and combinations to generate 80 billion Aadhaar numbers so that we are never in danger of running out of numbers to assign to people. To further strengthen the numbering scheme, a number once assigned to a person will never be reassigned, even after their death. There was some pressure for 'VIP numbers'—numbers that are seen as numerologically favourable, for example, or special numbers for ministers. The UIDAI disregarded all such requests so that everyone got one democratically assigned random number irrespective of who they were.

Reducing the technology footprint

A key design decision that the UIDAI's technology team took very early on was that biometric data would be used to ensure an Aadhaar number's uniqueness. We knew that every time a new enrolment took place, an individual's biometric data would have to be compared against every other data set already in the system to verify that this person had never enrolled before—a process called de-duplication. But could this be done 1.2 billion times? To find out, Raj Mashruwala flew to San Francisco to meet leading experts in the field of biometric identification.

Nandan recalls the aftermath of that discussion. 'I got a call from Srikanth Nadhamuni who was on his way to Dehradun to give a talk. He was in a bit of a panic as he said, the experts more or less think the way we're planning to do biometric de-duplication is going to be fantastically expensive, with no guarantee that it will even work.' In their opinion, the kind of heavy computational processing that de-duplication would require, to say nothing of the biometric scanners themselves, would drive costs to an unsustainable high. Nandan recalls, 'I told Srikanth that the UIDAI had already promised to deliver 600 million Aadhaars in five years and we could not go back on that commitment. The team had to find a way to make it work.'

Srikanth recalls experts opining that the team would end up with several cricket fields' worth of data centres to handle the enormous volumes of data the project would generate. Pramod Varma, UIDAI's chief architect, tells us, 'I remember being told that we would need a thousand computers worth a million dollars each for the de-duplication process, which is an expenditure of a billion dollars just to buy computing power.' What could the technology team do to bring costs down to a manageable level?

One option was to look at what some of the leading technology companies in the world were doing. As early as a decade ago, working at these scales would have seemed improbable enough to belong to the realms of science fiction. Yet, today we exchange over 100 billion emails a day;[6] Google handles 3.5 billion search queries daily;[7] there are currently over 1.28 billion people in the world using Facebook,[8] and over half a billion using WhatsApp[9]. Companies that have used the power of the internet to build new technologies and tools have become trailblazers in the art of working with a huge user base. The innovations they used to make this possible gave the UIDAI both the tools, and more importantly, the courage to forge ahead by taking calculated risks.

Another decision was to use commodity hardware—the same sort of computers that are freely available in the open market, and which aren't terribly powerful by themselves. But hooking up lots of them in parallel increased their computing power significantly, enough to tackle the de-duplication exercise. This system was cheap and flexible, avoiding any issues of vendor lock-in while being easily scalable and compact—the final data centre taking up only 10,000 sq. ft. rather than entire cricket fields.

A major technology challenge was to figure out in what language the actual enrolment would be carried out. Sanjay Jain, UIDAI's chief product manager and formerly at Google (where he co-authored the popular MapMaker tool) explains that 'urban residents tend to be more comfortable with English than their rural counterparts, who largely communicate in the local language of the state. Hence, data was captured both in English and the local language of the enrolment area. As the operator keyed in the data in English, the software automatically translated it into the local language, which

AADHAAR **ENROLMENT PROCESS**

Visits closest Aadhaar enrolment station

Provides basic demographic details: name, address, date of birth, gender

Provides biometric details: photo, fingerprints, iris scan

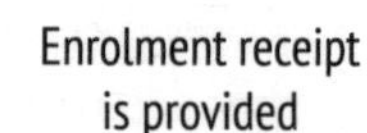

Enrolment receipt is provided

Aadhaar letter is printed and delivered by post to the resident's address

DATA REACHES AADHAAR SERVER

Demographic details are checked against database for duplicate

Biometric details are checked against database for duplicate

Aadhaar number is assigned if verification is cleared

70 REGISTRARS

300 ENROLMENT AGENCIES

120 CRORE RESIDENTS

could then be corrected for phonetic and spelling errors by both the operator and the enrolee.' The entire enrolment process is explained in the diagram on the facing page.

The unique heart of Aadhaar—biometrics

The technology team at Aadhaar was faced with the challenge of using biometric data at scales beyond anything the world had seen so far. In order to enrol 1.2 billion people, every person's biometrics would have to be compared with every other biometric in the system to ensure uniqueness. That's 700,000,000,000,000,000 (700 million billion) biometric comparisons. Storing all this information would generate 15 petabytes of data.

Hoping to learn from others' experiences, they studied other biometrics-based efforts, such as the US VISIT border security programme. Our team found that this programme was 'locked-in' with one vendor, forced to use only that vendor's software and biometric devices. At the time the team met with them, they were handling numbers on the order of 100 million identities. India is ten times bigger; what if we selected one vendor for Aadhaar and found that they were unable to work at this scale? Around the world, we'd seen that multiple government projects failed because they were helplessly yoked to a single vendor, who was unable to deliver for a variety of reasons. We could not afford to take that risk, and so we decided to use multiple vendors, both to provide the biometric scanning devices as well as to carry out the de-duplication process before handing out a new Aadhaar number. This approach had never been tried before, but by virtue of being the largest customer in this field, the government was able to dictate its terms. Now the UIDAI could never be held hostage by one supplier, and the power of the markets ended up driving volumes up and prices down.

Jagadish Babu, on a two-year sabbatical from Intel, managed the biometric device ecosystem in the early days of Aadhaar. He tells us that 'Aadhaar has single-handedly created a large market in biometric sensors and has spurred innovation in sensor technology worldwide. The demand generated by the UIDAI has led to a ten-fold drop in sensor costs, from over $5000 to $500 per device.' Biometrics used to be a restricted-use technology for security applications; now, it has been thrown open to the Indian public for the targeted delivery of social services—consider that Apple launched its fingerprint-based Touch ID on the iPhone in 2013, four years after Aadhaar was up and running. As a result of competition and scale, it was estimated that UIDAI's cost per de-duplication was the lowest in the world by a factor of three. Srikanth Nadhamuni explains the numbers to us. 'The worldwide benchmark for a single de-duplication was on the order of 31 cents (Rs 20). We managed to carry out de-duplications at a cost of 4 cents (Rs 2.75) per query.'

As enrolments scaled up, the team started finding aberrations in the biometric data due to mistakes in scanning. Vivek Raghavan, an expert in integrated circuit design, was one of the lead volunteers heading the UIDAI's early biometric efforts and the proud holder of what he calls 'the most interesting title at UIDAI'—his official title didn't even use the word 'volunteer', it just said Biometrics. He tells us, 'There were a number of data packets where the iris data for the left and right eye had been switched—it turns out the camera used for the iris capture had been inadvertently flipped. We found out about this error in early January 2011 and fixed it within a month, in which time we had to invent a bunch of algorithms to detect flipped iris images and correct them—you actually need to check the position of the tear gland in every image.' They caught these and similar errors only because they had a strong set of quality checks in place, allowing them to fix these mistakes quickly without forcing individuals to re-enrol and ensuring that all the captured data met quality standards.

Completing the cycle: The Aadhaar letter

Once enrolment and de-duplication are complete and an Aadhaar number is issued, every resident receives a printed letter with their number and demographic information. Today, that thick, laminated sheet of A5 paper is a familiar sight at airports, in trains and at bank branches. When it came to getting these letters out to people, there was considerable discussion on whether to use a courier service. But at the end of the day, the only agency in the country with the reach and the staff to decipher such cryptic names and addresses as 'X, Y's son, who lives behind the old school', was India Post. The postman is a government employee, and his delivery of the letter is considered as an official verification of one's address that can be used for other government procedures. He is also the person most likely to know every nook and cranny of the locality in which he operates as well as the families who live there, granting him the facility of decoding complex or incomplete addresses.

After generating an Aadhaar, the data would be sent to India Post for printing. As Ashok Pal Singh recollects, 'India Post never took us seriously when we told them that in a short time frame, one million Aadhaars will be generated daily—perhaps thinking how a government programme could scale this fast.' Very soon, the printed letters piled up, and people were waiting for months to receive them. In order to add capacity, other printers were contracted and the backlog was addressed. Letters were also sorted by PIN code and delivered to India Post at different points across the country to ease the load on the postal network.

Despite these efforts, letters do sometimes fail to reach the intended recipient, perhaps because the address is incomplete, the recipient has moved or the letter has been lost in transit. Enter electronic, or e-Aadhaar. The resident can go to the UIDAI's website, enter their enrolment number, download a copy of their letter and simply print it out for themselves. This printout has the same standing as the official letter issued by the UIDAI, and was made possible by a design decision that we discuss in a later chapter—the decision that Aadhaar would be a number existing in multiple formats, not just a number on a card; it would not be tied to a specific physical form.

Proving your identity with Aadhaar

What could a person do with their Aadhaar number once they had it? Its most powerful application is that instead of being just another identification number, it can actually be used to verify your identity electronically—the system can be queried and will provide a 'Yes/No' answer if you ask: 'Is this Nandan Nilekani?' No other ID system in India can do that. Moreover, Aadhaar provides not one but three different types of authentication: demographic, which verifies a person's name, address and their date of birth; biometric, which verifies a person's fingerprint or iris data; and a one-time password system, similar to that used in online banking, where a password for one-time use is sent to a registered mobile phone.

For strong, banking-grade identity verification, one has to carry out multifactor authentication, verifying at least two pieces of information in two different ways. For example, to use an ATM to withdraw money, you need a debit card ('what you have') and a PIN ('what you know'). Strong authentication can be provided by the UIDAI with biometrics ('who you are') combined with the one-time password sent to the mobile phone ('what you have'). Today, 900 million people can suddenly prove their identities online, in real time, using their Aadhaar numbers. This makes it possible to build an entire new class of applications, especially within government, that are shown in the diagram on the facing page.

The platform that built Aadhaar

Over the last two chapters, we have seen that both the Aadhaar scheme and the technology platform upon which it was built are quite unlike anything else in government. Aadhaar is perhaps the only government identity which is entirely open-ended; it can be used for any service, public or private, that requires identity verification. In the same way, Aadhaar runs on an open technology platform, using open-source software and standard, open-market hardware. A small team executed the project, keeping all the strategic thinking in-house while outsourcing the actual execution phase. One simply cannot outsource thinking and design to someone else: one has to get their hands dirty. (This is the exact opposite of the approach government usually takes.) Once the UIDAI put the appropriate technology

THE **MULTIPLE USES** OF AADHAAR AUTHENTICATION

PDS

NREGA

Janani Suraksha Yojana

Sarva Shiksha Abhiyaan

FERTILIZER

KEROSENE

LPG

FOOD

WATER

ELECTRICITY

Driving Licence

Passport

Airport Entry

Tax Returns and Payments

Land Titles

Certificates, Documents

Bank account opening, buying financial products

Business correspondents and microATMs, buying SIM cards

Voter ID issuance, while voting

Attendance of government officials

standards in place, any vendor could provide the necessary services. By bringing multiple vendors on board, competition increased, costs dropped, and scale was achieved.

Many of the decisions the UIDAI took went against the accepted norm, and the team had to accept both the associated risk and the criticism that came with them. However, the Aadhaar programme has now conclusively proven that we need not look only to the Silicon Valleys of the world for cutting-edge innovation in technology. Government is an equally fertile environment in which to build such solutions, not as an end in itself but as a means to improve the lives of all 1.2 billion of our fellow Indians.

3

BANKING ON GOVERNMENT PAYMENTS

SIX HUNDRED THOUSAND villages in India have no banking facilities whatsoever. What do their residents have to do to get access to financial services? If you are Basudeb Pahan, you trek fifteen kilometres through the hilly jungles of Jharkhand to reach the nearest bank branch, losing a good deal of money in the process. All across rural India, people perform similar feats of endurance. The ordeal of Ram, a villager in Madhya Pradesh's Atariya district, is chronicled in the accompanying diagram. After walking for an hour and a half, he must catch a bus to cover the twenty kilometres remaining between him and the local bank. The bank has decreed that his wages under the government's rural employment scheme, MGNREGA, will be available to him only on Thursday and must be collected before 2.30 p.m., the bank's closing time. If he misses the deadline, he has to come back the following week. To collect the $8 (Rs 500) due to him, he loses one day's wages $1.4 (around Rs 91), has to pay 15 cents (Rs 10) for the bus fare, and give a cut of his salary to the moneylender, who charges additional interest. In all, Ram effectively loses a fifth of his money before he can collect it.

The problem assumes a slightly different dimension for the urban poor, who are surrounded by banks yet unable to access their services for want of identity documents. Lal Singh is a migrant labourer in New Delhi who lives in a slum and wants to send his savings to his

RAM'S THURSDAY **ORDEAL**

Ram's efforts to collect his NREGA dues

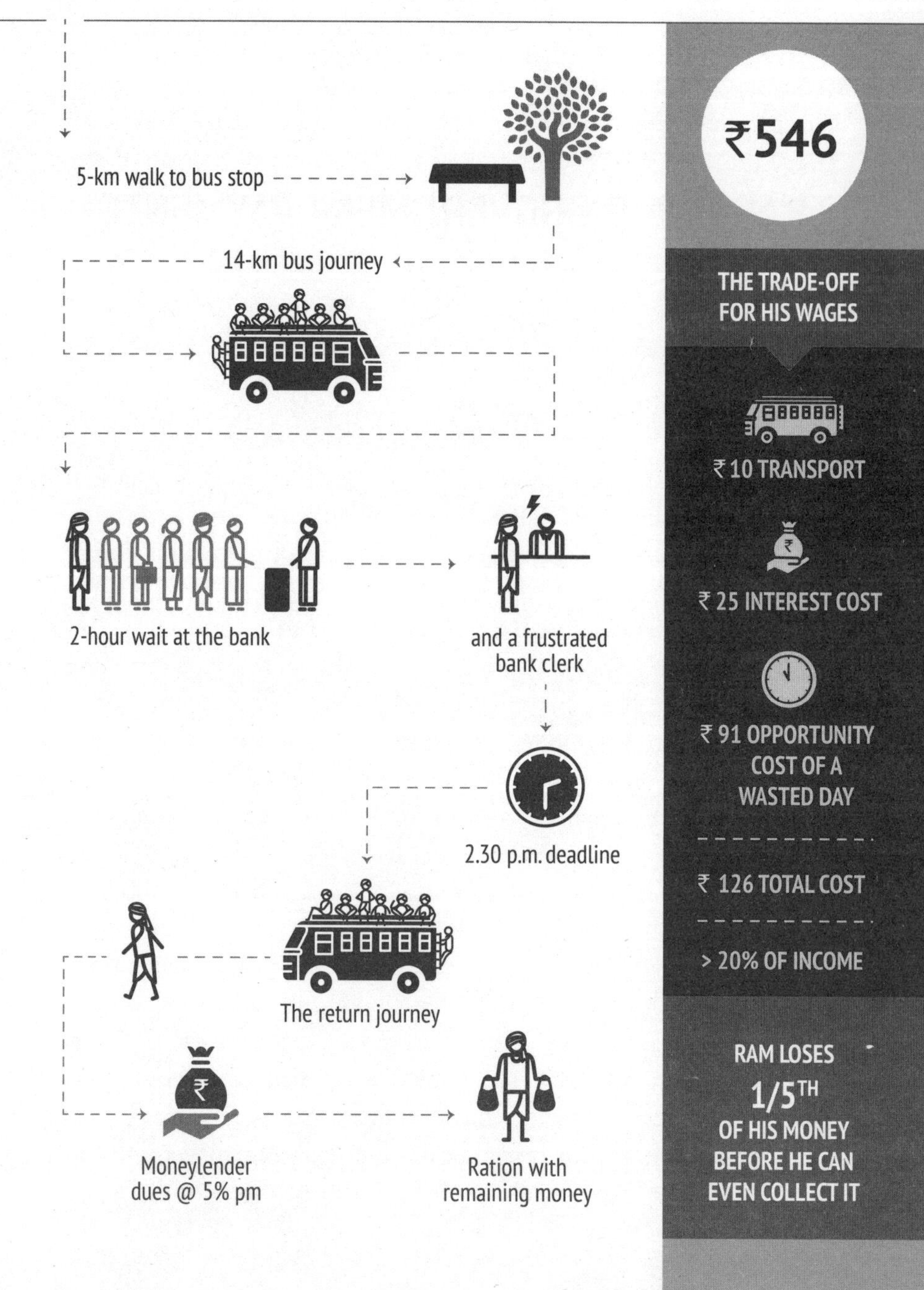

family back home in Bihar. Being a poor migrant, Lal Singh has no ID proof, and cannot open a bank account. What channels can he use? Initially, he used to send money through the post office, which charged him 5 per cent of the total amount as a service fee. His family complained that it took them a month or more to actually get the money he had sent. He then switched to a private agent, who would deliver the money directly to his family's home for a charge of 5–7 per cent of the total. However, the money still takes a few days to reach his village, and he has no way of checking that the money is being delivered correctly. A fellow migrant, Rashid Ul Sheikh, has no bank account and nowhere safe to store his money. The only ID document he has, a village ration card, is not accepted by the bank authorities in New Delhi. He is forced to hand over his savings to his landlord for safekeeping, who charges him 10 per cent for this service.

Basudeb Pahan, Ram, Lal Singh and Rashid Ul Sheikh represent the majority of Indians who find themselves excluded from the formal financial sector, and pay a high cost for it. Marginal farmers, landless labourers, the self-employed and those who work in the unorganized sector, urban slum dwellers, migrant workers, ethnic minorities and socially disadvantaged groups, senior citizens and women alike are unable to open bank accounts or access basic banking and credit services. An estimated 60 per cent of Indians, both urban and rural, do not have functional bank accounts. Out of India's 90 million farmer households, over 51 per cent have no access to any form of credit at all, while 73 per cent have no access to formal sources of credit. Other measures of engagement with the financial sector are equally dismal.[1]

The absence of a universally accepted identification has in a sense given rise to two economies in India. On one side, people open bank accounts, buy insurance, take out loans and transfer money to their families with relative ease. On the other side of this divide, the poor and disadvantaged rely on a shadowy system of moneylenders and ad-hoc financial services of questionable legality. Such basic activities as sending money to relatives become fraught with difficulties and costs. On a smaller scale, financial exclusion also sharply limits the ability of people to save, invest and improve their circumstances. Looking at the larger picture, this results in significant damage to India's economy. For example, only 3 per cent of Indians file income tax returns; in comparison, nearly 45 per cent of all US citizens pay federal taxes.[2] A nation cannot progress when a huge number of its citizens are cut off from accessing the fruits of the country's development.

The ultimate goal of financial inclusion would therefore be to bring banking and financial services to the doorstep of every Indian. The microATM that Basudeb Pahan uses to withdraw his pension now will be the instrument that helps us achieve this goal. The power of the microATM, combined with the ability to use an individual's Aadhaar data to verify their identity, is sufficient to transform the humble kirana store into a local bank branch in every village of India. The simplicity of this idea lends itself beautifully to building a large-scale banking network across the country, one that will usher every Indian into the formal financial sector.

Lurking in the shadows: India's parallel economy

Mumbai's Dharavi locality is a place of staggering contrasts. With its narrow byways and tumbledown tenements teeming with people, at first glance it lives up to its title of 'Asia's largest slum'. Beyond the clichés, however, Dharavi is an economic boom town, the beating heart of India's informal economy, generating anywhere from $600 million to $1 billion worth of output annually.[3] Viral was given a tour of Dharavi by the local branch manager of the Indian Overseas Bank. Despite having grown up in Mumbai, Viral had never set foot in the area; he was amazed by what he saw. Crammed into a warren of tiny rooms and sheds were factories churning out savoury *chivda* and other teatime snacks, workshops producing fine leather goods, an ice-cream manufacturing plant and a garment factory making T-shirts that were exported to Saudi Arabia and other countries—and this was just a sample.

What was the bank manager's perspective on the area? He explained that while many of these businesses were raking in handsome profits, almost none of them had a bank account—not surprising when you consider that 75 per cent of all Indian businesses are unlisted, and nearly 90 per cent of small businesses have no links at all with the formal financial sector. He was wrestling with the dilemma of offering loans and other financial products to companies that were otherwise solid bets except for one problem—on paper, they didn't exist.

India's informal sector generates 90 per cent of all employment in India[4]—yet these employees have no access to savings accounts, loans or insurance. The contributions of this 'shadow economy' to the nation's GDP are vastly underestimated.[5] The idea that boosting financial inclusion can have a huge positive impact on both the nation's economy and the everyday lives of its people is not a new one—the number of schemes launched to address this very issue bear testament to the fact. Clearly, there is still a long way to go; as recently as December 2013, the RBI governor, Raghuram Rajan, made a strong push for 'reach(ing) everyone, however remote or small, with financial services', and said that improved access would be one of the 'five pillars' on which further development would be based.[6] An RBI report also listed the six key vision statements that would help India achieve full financial inclusion by January 2016.

What happens to India's unbanked millions if they are given access to formal financial services? For one, their days of stowing money under the mattress will be over, and the usurious moneylender will become a figure of the past. They are already a finance-savvy population; in the book *Portfolios of the Poor*, which examines how villagers and slum dwellers in Bangladesh, India and South Africa manage their money, the authors found that the average poor household had 'a fistful of financial relationships on the go', demonstrating a surprising appetite for employing financial tools to manage their budgets while relying on informal networks and family ties to save money and take out loans.[7] Clearly, this appetite is not being satisfied by existing forms of financial access. And yet other sectors continue to successfully engage with this same population; even in some of India's remotest villages, you can go to the local cigarette and sweet shop and top up your mobile's talktime for as little as five rupees. If the telecommunications industry has worked out how to deliver services at high volume and low cost, overcoming geographical constraints along the way, why can't the banking sector do the same?

However, the strongest push for financial inclusion will not be from banks or other financial institutions. It must necessarily come from the government, which runs an enormous financial network disbursing an estimated $46B (Rs 3 trillion)—3.5 per cent of India's GDP—as social security benefits to the poor.[8] The bulk of this money is spent on schemes that guarantee employment to the rural poor (MGNREGA), offer scholarships to universalize elementary school education (Sarva Shiksha Abhiyaan, or the Education for All Movement), provide cash assistance to expectant mothers and health workers (Janani Suraksha Yojana, or Maternal Safety Scheme) and provide pensions to the elderly, widows and the disabled (National Social Assistance Programme).

How efficient are these social security schemes at delivering financial benefit to the poor? Two volunteers at UIDAI, Naman Pugalia and Mihir Sheth, travelled extensively across India to find the answer to this question. They learnt of leakages in the system that diverted resources away from desperately poor people, who then suffered the added insult of having to pay for services which should have been freely available to them. Minu, who was expecting a child, paid $10 (Rs 630) in order to receive a payment of $22 (Rs 1400) under the Janani Suraksha Yojana. Anjana Namdev received her $3 (200-rupee) monthly pension only once every few months. Suman Kadam had to depend upon the whims and fancies of the local postman for her payment to be delivered. Minzi was forced to mortgage her pension card for a loan to make ends meet. There were also other kinds of errors: some people receiving pensions turned out to be far too young to be eligible; in other cases, dead people's names were being used to illegally draw monthly payments. In every village Naman and Mihir visited, they found countless errors of inclusion and exclusion, a sad litany that explained why even the most well intentioned of government schemes failed to bring succour to the needy.[9] The cost of these inefficiencies, errors and leakages adds up to a staggering trillion rupees—nearly a third of the government's entire expenditure on welfare schemes.[10]

Every government payment that does not reach the intended

beneficiary, every government entitlement that a deserving individual is unable to claim, every error of inclusion and exclusion is in effect a broken promise made by the Government of India to its people. Millions of such promises are broken every year, even as the government continues to launch additional schemes, pump more money into the system, and enact new laws to ensure compliance and prevent fraud. The only way to ensure that these promises are actually kept is to build a new class of institutions that use technology to design solutions that work for everyone.

Bridging the last mile with electronic payments

Whether it is bringing banking to the masses or the disbursement of government benefits, the existing systems are clearly unable to keep pace with the needs of India's population. In some cases, the reasons are purely financial—it is expensive to build, staff and operate a new bank branch, and the kind of high-volume, low-value transactions that take place in a rural economy might not offset these costs. As a result, people are denied access to the financial services they need most: credit, so that they can borrow in bad times; savings and investment products, so that they can save in good times; and insurance, so that they can protect themselves against unforeseen circumstances and acts of nature—an accident, an illness, a crop failure or a flood. If these financial products are to be made available to everyone, the prices of such products must come down, and the associated risks must be offset by spreading them across the entire population.

Whether it's the printing press or the personal computer, technology has been the skeleton key for enabling access, the great leveller when it comes to making things cheaper and easily available. The ubiquity of the mobile phone in India is a textbook illustration. Why shouldn't this be true of the financial sector as well? In most financial transactions, it is in the last mile—where systems finally meet consumers—that things begin to break down. Here, costs are the problem. We've mentioned that the cost of creating a physical infrastructure is one deterrent. Another is the fact that cheques and cash are expensive to manage.

A significant amount of time and effort is expended in shepherding them through the system and finally into the recipient's hands.

We firmly believe that the only way to bridge the last-mile gap will be through the widespread adoption of electronic payment systems. The government must be the initial driver, using the heft and reach of its social security schemes to drive the adoption of an electronic payments model. As momentum grows, private players can step in. We envision electronic payments as the first step on the ladder of financial inclusion. The immediate next step would be the creation of bank accounts to house these payments. As soon as people have bank accounts, they become eligible for other financial services as well. The next rungs of the ladder are loans, insurance schemes of all kinds—for crops, health, life, accidents—and pension schemes. As people move upward, they become integrated into India's formal financial sector, with all its attendant benefits.

Electronic payments will also help to plug the leaks in government disbursement systems that drain precious resources away from the intended recipients. Once money is transferred directly into a beneficiary's bank account, the entire process becomes transparent. Payments can be easily traced and collected, and corruption will automatically drop, so people will no longer have to pay to collect what is rightfully theirs. The estimated savings from such a model are enough to boost the country's welfare spending by 25 per cent, or increase the per capita income of every poor household in India by 15–20 per cent.

In his book *The End of Money*, the journalist David Wolman narrates a story about the unexpected consequences of switching to an electronic payments system.[11] In 2010, local police in the Afghan province of Wardak started receiving their monthly salary electronically. When they checked their bank balances on their cellphones, they were stunned to find that their salaries had jumped 30 per cent overnight. This was not an accounting error or a sudden outburst of generosity on the part of the Afghan authorities—the extra 30 per cent was the amount that, unbeknownst to the policemen, corrupt senior officials in the system had been skimming from their cash payments.

For an electronic payments system to work, two major requirements are the ability to identify a beneficiary correctly, and a way to ensure that the money is reaching only the intended recipient. This is where Aadhaar enters the picture. Once the Aadhaar number is added to the database of all social welfare recipients, the government can now accurately identify these individuals. On the other side, linking a person's bank account with their Aadhaar number makes it possible to reliably verify the identity of the account holder. Aadhaar now serves as a link between the government and the people, making it easy both for the authorities to transfer payments to the correct individual's bank account, as well as for people to easily withdraw money using Aadhaar to authenticate their identity.

Many of these ideas have found a home in the recently launched Pradhan Mantri Jan Dhan Yojana (PMJDY), which will ensure that every household gets access to a bank account, a debit card, and up to $77 (Rs 5000) overdraft facility from the bank. Eventually, bank accounts under the PMJDY will become the repository of all government payments. As of August 2015, over 170 million bank accounts have been opened under this scheme.[12]

Breaking our addiction to cash: Why nations are going cashless

India is the fourth largest user of cash in the world. What makes cash-based transactions such an attractive prospect for most Indians? For one, there are no extra transaction costs involved when you pay with cash, costs that often make it financially unviable for smaller merchants to switch to electronic payments. While the local supermarket may have enough transaction volume to simply accept these fees as the cost of doing business, the owner of a corner stall selling bananas, shampoo sachets and cigarettes by the stick certainly doesn't have any incentive to allow customers to pay by card. A cash transaction is immediate, as simple as a banknote moving from one hand to another. You don't have to worry about a computer system crashing and losing your transaction. A thousand rupees loaded into Ola's digital wallet allows you to only buy taxi rides; the same thousand rupees as a banknote in your pocket can buy you anything under the sun. On the flip side, it's also easier to launder money and evade paying taxes if you're using cash, since those transactions are much harder to trace.

Part of the reason is also the fact that online payment systems in India still have very low rates of penetration. Large e-commerce companies have developed the workaround solution of cash on delivery (COD), much as mobile telephone companies came up with the idea of offering prepaid connections to customers whose lack of credit history disqualified them from buying the standard post-paid mobile connection.

Financial inclusion is one very good reason to shift from a cash-based system to one that operates on electronic payments, but there are other compelling reasons as well. Cash is a very expensive habit for the nation to cultivate. The cost of printing, managing and moving money around the country is as huge. In the period 2010–2011, the RBI spent nearly $369B (Rs 24 billion) on printing money, and an additional $7M (Rs 455 million) on distributing that money nation-wide. Cash has other problems too—it can be lost or stolen, and a wet, torn or otherwise damaged banknote is not accepted by most businesses, making it valueless. Despite what we may have seen in the movies, it's not easy to transport large amounts of cash, or move cash over large distances. When rampant hyperinflation in Zimbabwe caused an astronomical increase in the price of daily staples (a loaf of bread cost over 100,000 Zimbabwean dollars), hapless citizens had to haul money around in sacks and suitcases to pay for their groceries.[13]

So long as money circulates outside the formal financial system in the form of cash, the monetary policy of the government has no real effect on economic activity. Electronic payments, on the other hand, can greatly reduce friction in the economy, since transactions are simpler, faster and easier to trace. If the government were to make the switch to electronic payments, the savings in one year alone would be enough to pay for the entire cost of setting up the system.

The global retail sector has embraced the convenience and transparency of electronic payments. In Sweden, you can buy a magazine from a homeless vendor, or make a donation in a church, and pay with your credit card.[14] PayPal, was an early pioneer in online payments, and services like Square, Paytm, MobiKwik, Apple Pay and Ezetap (founded by our former UIDAI colleague Sanjay Swamy) offer a multitude of electronic payment options to the consumer, whether it's turning your phone into a mobile terminal that can process credit card payments, or letting you pay by simply tapping your phone against a point-of-sale terminal in a store. Smartphones like Apple's iPhone 6 now come equipped with fingerprint readers so that adding your fingerprint data to every online transaction increases the security and reliability of such payment systems. Taking things further, we have systems like Bitcoin that operate on 'digital currency'—not issued or controlled by any government—where users can transfer money completely free of regulatory oversight.

Lest one think that all of these innovations are confined to developed nations, consider M-Pesa, a mobile payment system operational in Kenya and Tanzania that's run by Safaricom and Vodacom, subsidiaries of the telecom giant Vodafone. M-Pesa has its roots in the airtime-swapping system that mobile phone users in some African countries spontaneously came up with; they started transferring mobile talktime minutes as a proxy for money transfers. Today, while only 7 million Kenyans have bank accounts, 18 million use M-Pesa; it processes a staggering US$ 1.6 billion in transactions every month. Data released by the Central Bank of Kenya shows a 10.6 per cent increase in formal financial inclusion for the time period 2009–13; it is not unreasonable to suppose that a significant fraction of that percentage has been driven by M-Pesa.[15]

Widening the financial net with Aadhaar

Dilip Asbe, now chief operating officer of the National Payments Corporation of India (NPCI), is possessed of a no-nonsense, 'get it done' attitude leavened with the bluff, hearty humour typical of his home city, Mumbai. As part of Viral's work on financial inclusion, he was scheduled to meet with Asbe. The meeting never happened, and

Viral assumed he had been given the kind of polite brush-off common in government circles. The real reason behind the cancellation? A motorcycle had run over Asbe's leg, leaving him limping for months. Viral recalls laughingly, 'Dilip helped to build Aadhaar-based payment systems with a limp and a smile!'

It was these and many other strong partnerships forged between the UIDAI and other organizations that helped to realize the vision of Aadhaar as the cornerstone of a new, technology-centric government payments system. One of the earliest organizations to come on board was the RBI, which had been active in promoting financial inclusion, including the use of handheld biometric devices to conduct secure banking transactions. The RBI had also pioneered the concept of 'business correspondents', external agents who were authorized by banks to offer certain services, such as cash withdrawals, in places where the bank did not have a branch.[16] Usha Thorat, former deputy governor of the RBI, tells us, 'When it was announced by the government that the UIDAI project was being launched, I felt it was like a dream coming true—providing the answer to the biggest challenge we faced for furthering universal inclusion. A unique identification based on biometrics was just what was required. I remember I was so excited I sent a message to Nandan saying that if we could link Aadhaar to financial inclusion it would be a winner.'

A key early meeting to chalk out the role of Aadhaar in electronic payments was held in December 2009; the attendees included representatives from the UIDAI, RBI, NPCI, all major banks and other related organizations. The final proposal integrated the RBI's business correspondent model with Aadhaar-based identity verification. We envisioned a system in which business correspondents would be equipped with a microATM device. Rather than a debit card or a PIN, a person would only need their Aadhaar number, their eyes and their fingers to identify themselves and carry out transactions instantaneously. This system was also designed to be interoperable—a customer of any bank could walk up to any business correspondent and withdraw money, in the same way that a Bank of India customer, for instance, can withdraw money from an Axis Bank ATM.

The details of Aadhaar-based financial inclusion were captured in a document entitled 'From Exclusion to Inclusion with Micropayments' released by the UIDAI. This document served as a blueprint and guide for the implementation of Aadhaar-based payments over the next few years. Today, over 15,000 microATMs are deployed by various banks and India Post, and should the government have the will, this number can rise to 1 million installations going forward.

The NPCI, formally incorporated only a year before the Aadhaar programme was launched, played a major role in the development of Aadhaar-based payment systems for the retail sector. A.P. Hota, the CEO of NPCI, along with its board members, was closely involved in these efforts; in addition to Viral, A.P. Singh and Rajesh Bansal from the UIDAI were part of the project as well. As Dilip Asbe put it, the goal of this partnership was to 'build payment systems for both Bharat and India'. The concept of using Aadhaar as a 'financial address' to deliver payments accurately—exactly the way a postal address ensures that you get letters meant for you—was further developed, culminating in the creation of the Aadhaar Payments Bridge (APB) and the Aadhaar Enabled Payment Systems (AEPS), for handling government and retail payments respectively. We discuss these systems in greater detail in subsequent chapters. The success of these systems can be gauged by the fact that in the month of January 2015 alone, the NPCI processed nearly 50 million government payments through the APB system, for a total value of over $307M (Rs 20 billion).

An ATM in every village

How can we bring banking facilities to the 600,000 villages in India that have mostly had none? Imagine a network of 1 million microATMs, operated by business correspondents—the local postmaster, the owner of the village grocery store—functioning exactly like the mobile top-up points that have mushroomed all over the country. Those 600,000 villages fall under roughly 250,000 gram panchayats. So if we assign three microATM-equipped business correspondents to every gram panchayat and the remainder to urban areas, we

THE **ALTERNATIVE** MODEL

Convenient Cash Withdrawal

Walks to the nearest Business Correspondent (BC) with MicroATM

Biometric Authentication

Withdraws cash directly

WHAT HAPPENS BEHIND THE SCENES

Biometrics, money transaction instruction and AADHAAR number transmitted

BC's Bank Account

Credit

INTEROPERABLE SWITCH

Authentication

Data verified on Aadhaar server

Debit

Authorization

Ram's Bank Account

ENTIRE TRANSACTION IN A COUPLE OF MINUTES

INSTANTANEOUS

SIMPLE

EFFICIENT

CHEAP

UBIQUITOUS

1 million business correspondents are needed to serve India, similar to the telco top-up network.

There are over **600,000** villages in India, with over **500,000** with no access to banking facilities.

would in effect have brought a bank to every doorstep in the country.

How would such a system work on the ground? The accompanying diagram outlines the process. The business correspondent is equipped with a microATM that also incorporates a biometric scanner, capable of reading fingerprints and performing iris scans. When a person wishes to withdraw money from their account, they go to the correspondent, who captures the customer's Aadhaar number, biometric information and the payment details, and sends it to the appropriate bank. Once the customer's identity is successfully verified, their account is debited, the correspondent's account is credited, and the cash is handed over to the customer. The entire transaction is completed within a matter of minutes; the whole process is simple, efficient and cheap, and seems like child's play compared to the hapless Ram's travails to collect his wages.

How did integrating Aadhaar into the business correspondent model make it better than what was already out there? Existing solutions required that an individual's fingerprints and banking data be stored on a smart card; creating and delivering these cards was an expensive process for banks. The data stored on the smart card was not available anywhere else, which meant that the person could not operate their account from a physical bank branch, could not use an ATM, and could not do business with a correspondent from any bank other than their own, unwittingly creating a monopoly in the market. This monopoly was worsened by the fact that employees of for-profit companies—for example, mobile service providers—were not allowed to function as business correspondents.

The Aadhaar-based model overcame these drawbacks in a number of ways. Many banks functioned as registrars of the UIDAI, enrolling people for Aadhaar; they could set up bank accounts for these individuals at the same time. Identity verification through Aadhaar is, by design, completed online, without requiring a card or any other physical document. If the customer's bank details were also easily accessible online—much in the way core banking systems operate—all transactions would be entirely digital, completely obviating the need for physical documents.

Online access conferred other benefits as well. Unlike the earlier smart-card system, customers now had the freedom to operate their accounts either through a business correspondent or through a physical bank branch or ATM. It was now possible to build an interoperable system, one where a customer could do business with any business correspondent from any bank. This would create a level playing field in which customers had more bargaining power, and monopolies could not arise. Thanks to the core banking system, a bank account opened in Mumbai can be operated seamlessly in Agartala, in effect following the user wherever they go; the microATM network was designed to confer the same mobility upon its users.

Finally, the UIDAI and NPCI worked closely with the RBI to widen the regulations around who could function as a business correspondent so that a greater number of people could now offer banking services in the field. The RBI finally agreed that profit-making companies would also be permitted to function as business correspondents. The official notification to this effect was released by the RBI on 28 September 2010—coincidentally, just a day before Ranjana Sonawane of Tembhli village became the first Indian resident to enrol for an Aadhaar number. These three principles, of online access, interoperability and the incorporation of for-profit organizations into the business correspondent network, guided much of UIDAI's thinking in the later design of the microATM system.

Going live: Electronic payments in the field

It was 7 p.m. in the dead of winter in Ranchi, a city of nearly 900,000 and the capital of Jharkhand. With less than twenty-four hours to go for the first field test of an Aadhaar-enabled electronic payments system, Rajesh Bansal, the assistant director general in charge of financial inclusion at the UIDAI, was told that the microATM device to be used in the test was not working. He recalls, 'I informed the team that nobody was going home until the problem was fixed, and made the necessary changes to the software myself.'

It had been a long road to get to even this point. Despite being what Viral calls 'an expert wielder of both the carrot and the stick', Rajesh had found it difficult to get things working from New Delhi. He says, 'I remember telling Viral outside the UIDAI office in Delhi that I was going to Jharkhand myself and wouldn't come back until I got everything sorted out and the first live transaction went through successfully.' Once he got there, he had to deal with one challenge after another. The payments were supposed to be disbursed under the MGNREGA scheme; by the time a specific block and panchayat of Ranchi district had been identified as the test location, the MGNREGA work for the season had already been completed. Rajesh had to go to the district commissioner and request him to restart the project in that area.

The work was completed on 22 December, with payments scheduled to go live two days later. Rajesh spent most of the 23rd in the block development officer's room trying to get the payments released for distribution, drinking endless cups of tea and refusing to budge until he had the paperwork in his hands. That evening, while the microATM snafu was being sorted out, Rajesh also had to ensure that all the payees had bank accounts, and he furnished a personal guarantee to the bank (Bank of India, in this case) that he would provide the required documentation for those who didn't.

The first payment was scheduled for 9.30 a.m. on the 24th. At 8.30, the bank again informed Rajesh and his team that there was a problem of some kind, which they managed to fix. Rajesh says, 'Until 9.20, nothing was happening and we were getting seriously jittery. Finally at 9.24, the first account got credited electronically. That was an amazing moment. I messaged Nandan telling him that his vision had finally come true in the field. The villagers couldn't believe it either. It used to sometimes take them six months to get their money and now they were getting it in a matter of minutes.'

The groundwork for this pilot project had begun three years ago, within the few months of the UIDAI itself being established. Nearly a year before the first Aadhaar number would be issued, we were already designing a payments network based on Aadhaar and biometric authentication. There were two reasons for getting such an early start;

one was that we were giving ourselves enough time to iron out the kinks in the design and build a working solution that would be ready to hit the ground as soon as Aadhaar enrolments gained momentum. The second was that building such a network also required us to build consensus across all stakeholders—banks, payment industries, central and state governments—which we knew would be a lengthy process.

The same principle of asynchronicity that guided the design of the Aadhaar platform also informed our decision to start building both the system and its future applications simultaneously. We believe that asynchronous design should be adopted widely in all government projects—too often, waiting for one part of the system to be ready before beginning work on the next leads to bottlenecks and delays, a state of affairs that we were largely able to avoid.

What were some of the factors we had to consider when drawing up the blueprint of the electronic payments system? Firstly, it would have to reach millions of people, and the only network in India with the kind of reach and ubiquity that was required was the mobile telecommunications network. The mobile telecom network would have to be the first piece of the puzzle. Secondly, the system had to be based on a technology that would be easy to use for poorly educated or even illiterate people, who might not be able to remember or enter a PIN number into a device. Thirdly, the system had to ensure that social security payments were being correctly targeted to their intended beneficiaries, while also raising the barriers against potential fraud. This meant that any identification mechanism would have to be extremely specific to the person using it so that nobody else could fraudulently claim their benefits. With these constraints, the only technology that answered all requirements while still allowing deployment at large scales was that of microATMs using biometric authentication.

The UIDAI set up a microATM committee chaired by Prof. H. Krishnamurthy from the Indian Institute of Science, which worked for over three years and met more than twenty times to firm up the microATM standards. Such a long period of deliberation is rather uncharacteristic, but the committee was determined to produce a working solution rather than just a set of guidelines. The entire system was designed to function as a thin layer on top of the existing banking infrastructure. It was compatible with existing industry regulations and standards so that it could be easily adopted for use without much effort on the part of banks or payment corporations. The microATM itself would use Aadhaar-based biometric authentication to identify the customer, who could then carry out withdrawals, deposits, transfers and balance inquiries.

While the microATM technology was being put in place, the question of incentives had to be figured out. In a payment transaction, the party that benefits the most from that electronic payment typically bears the cost. In this case, the clear beneficiary was the government, which would profit from the high levels of transparency and accountability provided by the Aadhaar-linked microATM network. The savings generated by switching to this model were sufficient to cover the government's transaction costs; while the official recommendation was 3.14 per cent, the government currently pays a transaction fee of 2 per cent.

The operational model for microATMs is much like the public call office (PCO) model, where thousands of small entrepreneurs benefited from operating a pay-per-use phone. The only capital investment needed is $231 (Rs 15,000) to purchase a compact operating device, much like a payment card terminal. Factoring in the cost per transaction 23 cents (a maximum of Rs 15), the total cost is far lower compared to the expense of setting up and operating an ATM or a rural bank branch, making the microATM model financially attractive to banks.

In the first few years after implementation of this model, we expect that cash will still be moving across the country, since people will withdraw the money disbursed into their bank accounts for their own consumption. There are many benefits to this process: money is now transferred to a bank account without requiring any intermediaries or approvals. As the following diagram shows, an Aadhaar-enabled bank account should ultimately allow the customer access to funds from multiple sources, whether they be a salary paid by a private company or disbursements from government social security schemes.

BENEFITS OF AN AADHAAR-ENABLED BANK ACCOUNT

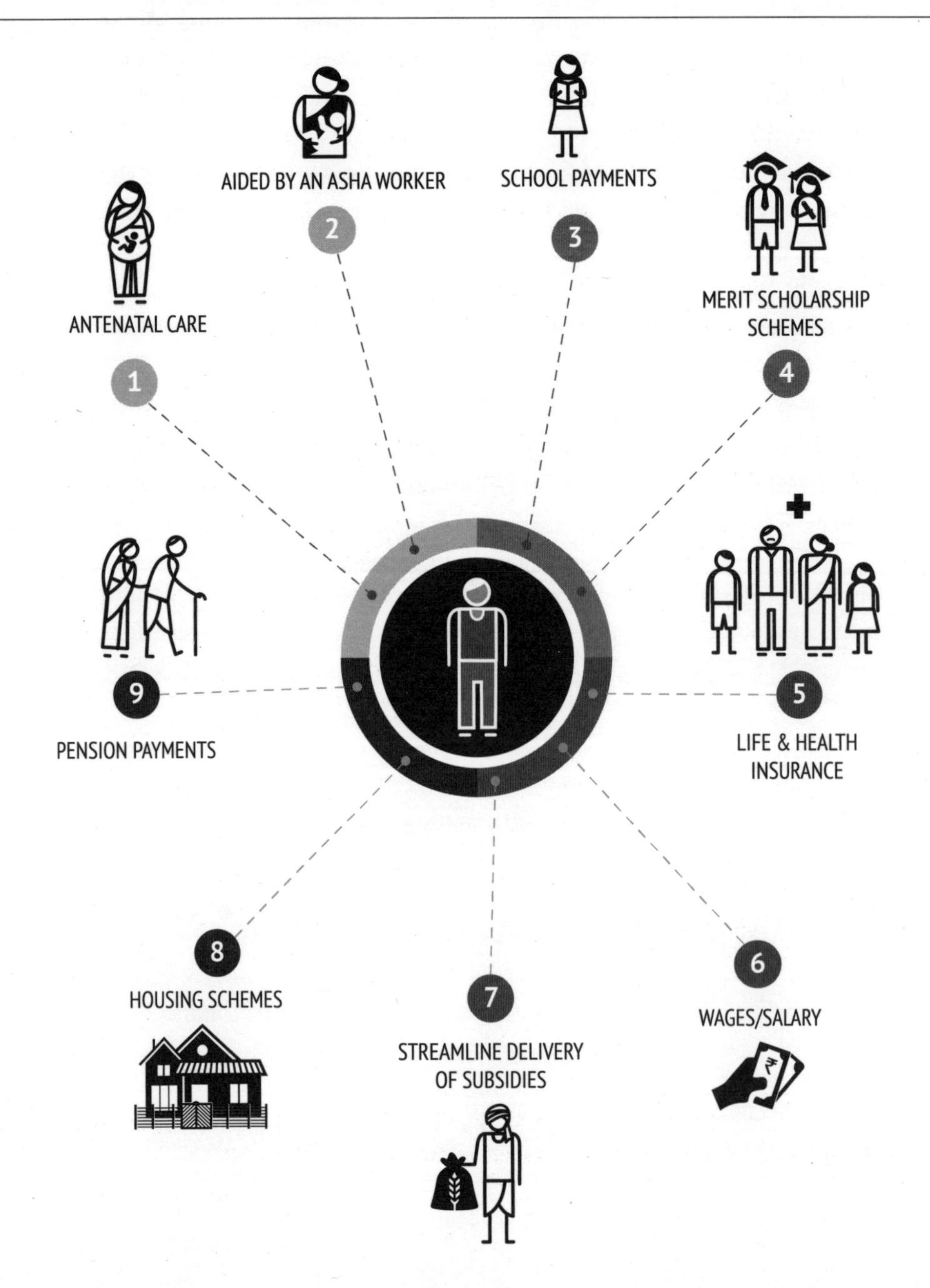

The prescription for a cashless India

In the accompanying diagram, we sketch out a road map for the transition to a cashless economy in India. As we will discuss in the next chapter, Aadhaar can serve as documentation to complete an electronic Know Your Customer process, turning what is currently a heavily paper-based system into one that is entirely digital. Aadhaar-based e-KYC has been used to open many of the 170 million bank accounts under the Jan Dhan Yojana scheme, and soon over $46B (Rs 3 trillion) worth of government social security and subsidy benefits will start flowing into these accounts. In order to handle this expanded customer base, a strong network of 1 million business correspondents will spring up, just like it did in the case of mobile phone operators providing top-up services.

We have already discussed some of the regulatory changes needed to support a move towards electronic payments: the business correspondent model and interoperability among banks, among others. The RBI has also further deregulated the banking sector, allowing small banks and payment banks to enter the arena.

But what about technology? Within a few years, one in every two Indians will own a smartphone, giving them access to the burgeoning ecosystem of online payment applications. We anticipate that the next year or two should see the launch of smartphones with iris readers, making it possible to use Aadhaar as an authentication factor for online transactions.

Another contribution towards a cashless society is the launch of a Unified Payments Interface (UPI) by the NPCI.[17] The goal of the UPI is to create a single, seamless platform for any kind of electronic transaction, whether it's payments from the government (subsidies, for example), payments to the government (taxes), payments from customers to vendors, and ultimately even money transfers between two individuals. The danger with so many players is that each will evolve their own payment method, making things unnecessarily complex. The UPI is designed to prevent such 'payment islands' from

1

e-KYC

Access to financial services through paperless KYC

2 **JAN DHAN**

Bank accounts for everyone

3

SUBSIDIES AND GOVERNMENT PAYMENTS

300 million beneficiaries get bank accounts

4

BUSINESS CORRESPONDENTS

1 million Business Correspondents bring banks to every doorstep

5 **REGULATORY INNOVATIONS**

Allow new business models to emerge

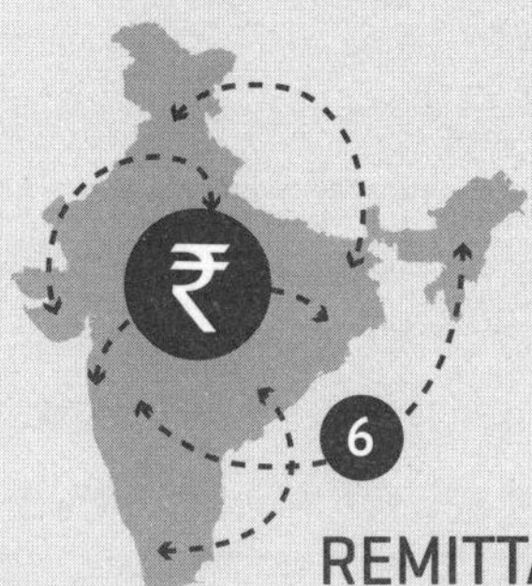

6

REMITTANCES

Electronic through Aadhaar, bank accounts and smartphones

7
MOBILE PAYMENTS
New systems for smartphone payments
8
ELECTRONIC MERCHANT PAYMENTS
Smartphones pay for everything from taxis to vegetables
9
SECURITY
Aadhaar biometric authentication
10
COST REDUCTION
Technology will cut transaction costs
11
PORTABILITY
Use any account to receive subsidy payments
12
SMARTPHONE APPS
New ways to process payments

emerging, and to create a central system that anyone can join and use. Since it is entirely digital and runs on existing infrastructure, the UPI is also cheap to implement. Nandan currently serves as honorary advisor (innovations and public policy) to the NPCI, focusing on the development of the UPI.

As more people start entering the electronic payments network, they start creating a digital trail of money through their transactions which can be analysed and, based upon their size and regularity, converted into a credit score. As a result, credit can now be offered to those people who, not being in the formal system, were not deemed credit-worthy, like the many enterprising small businessmen Viral met in Dharavi. Loans, for example, can now be disbursed directly through microATMs, with instant approval, and can even be paid back through microATMs. No longer do these loans have to be under the purview of banks; any financial institution capable of analysing digital transaction data—a person's 'data exhaust'—can make the decision to extend credit.

Building better government systems through technology

Ultimately, the key principles that drove the creation of the entire Aadhaar-based payments ecosystem are universal enough to be applied to any technology-enabled solution implemented on a large scale; many of these find an echo in the design of the Aadhaar programme itself. One is to build a simple thin solution, a platform that can easily be layered on top of existing payment systems with minimal effort. Customers must have the freedom of choice, in this case being able to carry out transactions with a representative from any bank through the creation of an interoperable payments network.

The economics of the system must be such that participants find it easier to make money legitimately rather than through corruption and rent-seeking. By creating standardized and simple processes, scaling up the network to include millions of customers becomes easier. Thanks to the 'network effect', each individual added to the system enriches it exponentially, creating a large enough user base that it expands under its own momentum.

And finally, all solutions must recognize and harness the energy of the Indian market. Today, we have emerged as a nation of entrepreneurs, a development that Nandan charted in *Imagining India*. We cannot afford to ignore this vast wealth of human capital. Entrepreneurs can drive innovation, while the strength of the government lies in creating infrastructure and the necessary regulation, as well as building scale. The Aadhaar-enabled payments network was designed to leverage both these strengths, using markets where possible and governments where essential, allowing each to focus on their core competence, all the while building a solution that worked for everyone.

With Aadhaar enrolments proceeding rapidly, we now have over 900 million residents equipped with an Aadhaar number ready to operate bank accounts and withdraw money; with the technology for microATMs in place, building a network of a million business correspondents to serve these potential customers is easily achievable. With the number of smartphone users expected to go up to over 500 million by 2015, we may soon see a shift to smart phone-based payments as well.

Creating demand for an electronic payment system might have been challenging, but when the government chooses to route even a small part of $46B (Rs 3 trillion) worth of social security payments through microATMs, the incentives arise automatically—the sheer scale of the numbers involved generates enough demand for the entire infrastructure to be built. Banks will set up microATM networks with a business correspondent in every village, vendors will compete on building microATM devices, and people will open accounts in a bank of their choice as multiple banks compete for their business.

The involvement of the government has converted what might have been a niche service into one that is freely available to all. Electronic payments were once considered a luxury for higher-end bank customers; now, they have become a utility that everyone can avail of. The basic banking infrastructure that will arise in every village of India will be the engine that powers further innovation in financial services for the benefit of every citizen; it will help create a truly integrated and inclusive economy.

4

MENDING OUR SOCIAL SAFETY NETS

> There are people in the world so hungry, that God cannot appear to them except in the form of bread.
>
> —Mahatma Gandhi

KRIPA SHANKAR IS a slight, moustachioed farmer from a small village near Amethi in the state of Uttar Pradesh. When Viral met him in 2012, he was pushing his bicycle by the side of a large highway, their conversation periodically drowned out by the ear-splitting honking of trucks as they roared past. A journalist, interviewing Kripa Shankar, pointed to a bag of fertilizer perched securely on his bicycle. Like farmers across India, Kripa Shankar had bought this fertilizer at a rate far below the market value, thanks to the generous subsidy provided to farmers by the Government of India. How long did it take him to get this fertilizer, asked the journalist. He replied, 'Twenty days! After twenty days of waiting for the fertilizer to arrive at the shop, there was a queue this morning and I waited for five–six hours. It's probably too late for my crops now but I had no choice.' How many bags did he need? 'Five.' And how many could he get? 'Two.' It was very likely that to get even those two bags, Kripa Shankar had to bribe the shop owner. What about his crops? 'Brother, they're ruined, come to my fields and see for yourself,' sputtered the indignant man.[1]

Kripa Shankar is not alone in his discontent. In 2008, farmers in Karnataka, upset when they received smaller quantities of chemical fertilizers than they were entitled to, vented their ire by burning a fertilizer shop, three buses and a police vehicle. Several more buses were damaged and thousands of farmers blocked the streets. Eventually, the police resorted to beating up the protesters and firing tear gas shells in an attempt to quell the violence.[2]

The broken disbursement system evoking such rage in India's farmers has its origins in our long and painful history of recurrent famines, reaching a nadir with the pre-Independence era famine in Bengal that caused an estimated 1.5 to 4 million deaths due to starvation, malnutrition and disease. Heavily dependent on the monsoon to provide irrigation for crops, and saddled with past policy failures that worsened the scarcity of essential foodgrains, an independent India resolved that improving national food security was a priority for building a strong nation. One path to this goal was to boost the nation's agricultural productivity. Agricultural research institutes were set up across the country to build a strong scientific infrastructure for agriculture. In a further attempt to break India's dependence on foreign food imports, eminent scientists such as the Nobel Laureate Dr Norman Borlaug and Dr M.S. Swaminathan, the statesman C. Subramaniam and others collaborated to build plant breeding programmes and irrigation development schemes that ushered in India's Green Revolution, allowing India to finally achieve self-sufficiency when it came to foodgrains.[3] The long-ranging success of this programme can be gauged by the fact that India is now neck and neck with China as the world's leading exporter of wheat and rice.[4]

As food production levels began to increase, the government took a series of policy decisions to ensure that agricultural productivity remained high, guaranteeing a steady food supply to the nation—decisions meant to protect and promote agriculture that exist to this day. Seeds are supplied at subsidized prices. The government has made substantial investments in large irrigation projects so that ample water can be provided at subsidized rates. Subsidized power is used to pump water for irrigation. Heavy subsidies on the manufacture and sale of fertilizers have made them both cheap and ubiquitous. Agricultural output also falls under this protectionist mantle; the government is one of the largest buyers of foodgrains and it sets the prices for procurement, assuring the farmer that there will always be a buyer for his crops. To some extent, these measures have been effective. India has weathered more recent episodes of poor rainfall and low crop yields without the masses descending into starvation.

Farmers are not the only ones to benefit from government subsidies. Around the world, governments provide subsidies for such commodities as food and fuel. India recently passed the National Food Security Act, designed to provide subsidized foodgrains to nearly 820 million people, almost two-thirds of the country's population.[5] Social security pensions, unemployment benefits and health and disability insurance are part of the safety nets created by governments to support the vulnerable and underprivileged.

All told, the Government of India annually spends nearly $46B (Rs 3 trillion) on schemes involving direct cash transfer and subsidies. That's almost 3.5 per cent of the nation's GDP—enough money to lift nearly every household in India over the poverty line. Over 100 million farmers benefit from agricultural subsidies, over 120 million households receive subsidized LPG, a cooking fuel, and a network of nearly 500,000 fair price shops dispense food and kerosene at subsidized rates to about 180 million families every year.[6]

As an attempt to eradicate social inequities and provide equal access to essential commodities, the inherent fair-mindedness of such systems cannot be denied. But, as Nandan has argued in *Imagining India*, the distribution system for subsidies is hopelessly broken. The government haemorrhages money while millions of Kripa Shankars across the country receive a fraction of the benefits they are entitled to. Grain rots in warehouses while the poor go hungry; although India has one of the biggest grain stockpiles in the world, and even exports some of it to countries like Saudi Arabia and Australia, one-fifth of the Indian population remains malnourished.[7] When these distribution systems fail, they hit precisely those people with the least resources to weather the storm on their own. Farmers in Karnataka set fire to buses; in Tunisia, a poor street-food vendor harassed by the police set fire to himself instead, triggering the Arab Spring revolutions in the process. While Mohamed Bouazizi was protesting police brutality, underlying the complex web of events that brought nations to the brink of civil war was a rise in food prices;[8] heavily dependent on foreign imports, these countries were deeply vulnerable to spikes in world food prices that the subsidies provided by their governments could not absorb.

The two major problems that have crippled public distribution networks are the inability to provide targeted services and a lack of transparency. Citizens lack both control and information at every step. Lists of intended beneficiaries are riddled with errors of inclusion and exclusion. Ghosts, fakes and duplicates are siphoning off supplies while corruption and regulatory procedures have raised the entry barriers so high that many of the truly deserving are left out in the cold. An opaque network means that the movement of goods from supplier to the consumer cannot be traced; stockpiling and hoarding of goods can neither be caught nor prevented. A lack of choice means that consumers are forced to purchase goods from a single supplier or fair price shop, creating monopolies in the market. A dual pricing system in which goods can be purchased at an artificially low price and subsequently sold at the prevailing market rate provides opportunities for arbitrage and market distortions. Analyses carried out as part of the 2015 Economic Survey reveal that rich households benefit disproportionately from multiple subsidies, while the poor for whom they are truly meant continue to struggle. The good intentions that birthed the idea of subsidies as a social safety net have long since been lost in a witches' brew of inefficiency, corruption and a blinkered world view that clings to outmoded systems past their shelf life.

Whether the government should be providing subsidies at all has been debated for years. At one end of the spectrum are those who feel the subsidy system should be dismantled altogether, and market forces should prevail. On the other end are those who feel the government is actually not doing enough, and needs to increase the reach and extent of its social welfare programmes. In our opinion, withdrawing subsidies and benefits altogether is too drastic a move—many of these are the outcomes of great political bargains in our society. Instead, we advocate a hybrid system that delivers benefits to the underprivileged while also being responsive to market forces. We envision technology as its lynchpin, allowing us to do more with less—better welfare programmes that place less of a financial burden on the state. Most importantly, technology allows us to deliver subsidies directly to the citizen; instead of trying to fix the broken disbursement system that has become the Achilles' heel of so many welfare programmes, we can eliminate it altogether.

Our country is in the grip of what economist Norman Myers refers to as 'perverse subsidies'—hurting not just our economy, but our environment as well.[9] While subsidy reform may have an immediate economic impact, in the long run it also offers us a chance to develop a new class of environment-friendly policies. As we over-fertilize our fields, deplete our water tables, chop down our forests, demand more coal and fossil fuels to power our industries and our vehicles, pollute our rivers and poison our air, it's a chance we cannot afford to let slip.

Good intentions, bad outcomes

The subsidy economy operates under a set of distorted rules that bear little relation to the rough-and-tumble world of the open market. In an open market, the scarcity of a particular commodity is signalled by a price rise and the market usually responds accordingly. Faced with an increase in onion prices, a housewife might simply choose to buy fewer onions or none at all until the prices stabilize. If fertilizers are expensive, farmers will use them sparingly. In the closed, artificial system created by subsidies, these market forces are no longer in effect, with harsh consequences.

Let's look at fertilizers first. Subsidies keep prices low, and farmers use them lavishly; according to World Bank data, India's usage of fertilizers outstrips that of the US.[10] This is a massive environmental problem, because fertilizer overuse—that of urea in particular—has the paradoxical effect of reducing soil quality and decreasing crop yields. While urea supplies plants with nitrogen, an essential nutrient, it must be added as part of a balanced fertilizer mix that includes two other essential nutrients, potassium and phosphorus; add too much of one and too little of the other, and crop growth is affected. Urea is a subsidized fertilizer, while others are not, which helps to explain the skewed balance of fertilizer usage. An agricultural development officer in the state of Punjab, traditionally considered India's breadbasket, complains that 'a farmer will become bankrupt, but he will not stop using urea', likening it to an addiction. Unable to comprehend why their crop yields are falling, farmers are dumping more urea onto their increasingly sickly-looking fields and unwittingly worsening the problem.[11]

Our dwindling water resources are a second concern. The government gives farmers power subsidies, making it cheap to pump water for irrigation. With little incentive to be prudent, farmers can afford to pump far more water than they actually need. Satellite measurements of groundwater tables have recorded a steady annual drop, and farmers now have to dig ever deeper bore wells to find water for their crops.[12] According to these studies, 54 per cent of India faces 'high' to 'extremely high' water stress, and groundwater levels in northern India are 'more critical than anywhere else on earth'.[13] The bulk of agricultural subsidies end up benefiting farmers with medium to large holdings who can afford to pay the market price; poor and landless labourers, the lowest on the agricultural totem pole, gain little or nothing from such policies. The 2015 Economic Survey further explores these misdirected subsidies—fertilizer subsidies, for example, largely benefit manufacturers rather than consumers. The poorest 20 per cent of Indians consume far less power than their richer counterparts and thus profit less from power subsidies. The bulk of water subsidies are allocated to private taps, whereas the public taps that most poor households draw their water from remain unsubsidized.[13]

Agricultural produce is also artificially insulated from the normal market rules of supply and demand. The government buys grain at a fixed price, benefiting the farmer but driving up costs for everyone else. The previous government budgeted a staggering $19 billion (over Rs 1 trillion), nearly half of its entire subsidy spend, on closing this price gap.[14] When farmers are given price support for crops like rice and wheat, they tend to over-cultivate these at the expense of other, non-supported crops. Agricultural patterns become skewed and the supply–demand equation goes out of whack, adding to price inflation.

When this closed system meets the open market, the consequences can be severe. Farmers used to obtaining a guaranteed price for their produce are ill-equipped to deal with the vagaries of the open market. They are further crippled by government policies designed to prevent hoarding and control inflation, such as banning exports and preventing farmers from selling their produce directly to private buyers. The rising price of foodgrains and vegetables is a perennial sore point, and the cost of the humble onion has become a rallying cry capable of bringing down governments. It is a sad testament to the failure of our agricultural policies that farmer suicides, driven by crop failure, debt and sheer desperation, continue to make headlines to the present day.

Perverse incentives operate in case of fuel subsidies as well. The subsidies for petrol and diesel were dismantled very recently, but their effects continue to linger. At its peak, diesel in India was a whopping 32 per cent cheaper than petrol, thanks to which 53 per cent of passenger vehicles sold in India over the 2014–15 fiscal year were diesel-powered, compared to a miniscule 3.2 per cent in the US.[15] It's still common to see luxury Mercedes sedans running on diesel in India. Just as common is the sight of a vehicle parked outside a restaurant as an employee unloads cylinder after cylinder of domestic LPG, obtained illegally because, of course, domestic LPG is subsidized while commercial LPG is not. The same domestic LPG also powers autorickshaws, since auto LPG is not subsidized either. There's a lot of money to be made in exploiting these loopholes; according to one estimate, 37 per cent of all domestic LPG consumers are 'ghosts', fake accounts created by buyers and distributors.[16]

Subsidized kerosene is provided as an affordable cooking fuel to the poor. Unfortunately, this has also given rise to the 'kerosene mafia', who illegally obtain subsidized kerosene and sell it for far higher prices on the black market. The system is so powerful and well entrenched that when Yashwant Sonawane, the additional collector of Malegaon in Maharashtra, allegedly tried to stop a local hoodlum stealing kerosene from a tanker, he was doused in the very same kerosene and burnt to death.[17] Earlier, in 2005, an Indian Oil Corporation manager named Shanmugam Manjunath had ordered a petrol pump near Lucknow to be sealed for selling fuel adulterated with subsidized kerosene; the petrol pump owner murdered him in retaliation.[18] These stories helped catapult the kerosene mafia into the public eye. Manjunath's death inspired a biopic, and Sonawane's murder was mentioned by Pranab Mukherjee, then the finance minister, in the Budget speech of 2011–12.

While the fuel subsidy burdened the Indian exchequer to the tune of $15 billion at its peak, far greater and much more worrisome is the insidious impact of such heavy fossil fuel usage on the environment. Recognizing the financial and environmental repercussions of subsidizing fossil fuels, governments across the world, from Jordan to Iran, are slashing subsidies and raising fuel prices. Research from the International Monetary Fund (IMF) shows that most fuel subsidies are misdirected; only 7 per cent of the subsidies in poor countries go to the poorest 20 per cent of households, while the richest 20 per cent of society claim a disproportionate 43 per cent.[19] Cutting these subsidies altogether would result in a worldwide savings of over $500 billion, accompanied by a 6 per cent drop in global carbon emissions by 2020.[20] A further impetus to the scrapping of fuel subsidies came from the G20 Pittsburgh Summit.[21] A declaration released by the G20 nations, a group which includes India, pledged to phase out 'inefficient fossil fuel subsidies', which 'encourage wasteful consumption, reduce our energy security, impede investment in clean energy sources and undermine efforts to deal with the threat of climate change'.

Technology as superglue: Mending a broken system

The central government currently offers subsidies on over a dozen commodities, including everything from coal to car parts, to oil, jute and cattle fodder; state governments can and do add to that list. This bouquet of subsidies doesn't come cheap; as per the 2014 Budget, the central government was projected to spend nearly $62B (Rs 4 trillion) on subsidies, over 4 per cent of India's GDP; in comparison, the country's defence budget is around 2.5 per cent.[22] That is a very heavy burden for an economy to bear.

Different goods are subsidized in different ways. In the case of fuel, for example, subsidies operate right down the production chain, starting with the purchase of raw materials and ending with the consumer. The subsidy burden eventually becomes the responsibility of state-run enterprises, leaving them grappling with heavy losses year after year. Fertilizer manufacturers, on the other hand, pay market price for raw material and production, and their expenses are later reimbursed by the government, showing up as excess expenditure on the government's balance sheet. As the prices of raw materials soared over the last decade, budget deficits and government debt grew likewise.

Subsidies are applied early in the production process of most commodities so that, as goods move through the supply chain, their prices are already low. In practice, this means that consumers can buy products at two prices, either the standard market rate or the cheaper subsidized rate. This price gap is ripe for exploitation by large, well-organized gangs like the kerosene mafia. Ordinary individuals exploit these loopholes as well; one enterprising maid buys subsidized rice from her home state of Tamil Nadu and sells it at market rates in Bengaluru, using her profits to fund her children's education. These are just a few of the countless examples in which the subsidy regime has turned out to be completely counterproductive, causing economic harm and actually working as a disincentive towards growth and development.

How does the government pay off its enormous subsidy spends? Its ability to do so is dependent on three factors: the international market price of commodities; how much money it can generate through revenue and borrowing; and the efficiency of its subsidy administration. The first factor is largely out of the government's control—it has no choice but to pay the going market rate for raw materials. The second factor, the state of the nation's treasure chest, is dependent upon the economy. We could raise taxes to generate more revenue for spending, but this approach runs the danger of stifling economic activity if taxes get too high. The only lever of control we realistically have as a society is to pay for subsidies by cleaning up our subsidy regime.

The first step in the clean-up is to recognize that our subsidies work at the wrong level in the system. Instead of being applied to manufacturers or other intermediaries, subsidies need to operate at the level of the consumer. As soon as we do this, the dual-pricing system for goods will vanish, removing all incentive for diversion and fraud. The advantages of such a direct benefits transfer system are detailed in the accompanying diagram.

Developing a direct transfer system for subsidies on LPG, kerosene and fertilizers was the mandate of a task force instituted by the then finance minister Pranab Mukherjee and chaired by Nandan. The task force released its recommendations as a report in 2011,[23] outlining a mechanism to ensure that subsidies were correctly targeted only to the truly deserving. What were some of the challenges that such a mechanism would need to overcome? Apart from the multiple opportunities for pilferage and fraud, the current model also robs consumers of their freedom to choose. The government specifies which goods are to be subsidized, and where you can buy them—subsidized wheat, for example, is available to you only at the ration shop in your locality and nowhere else—due to which unintended monopolies and distorted production patterns emerge. An ideal system would allow for people to choose when and where they wish to avail of subsidized goods, whether it's a physical product or a cash transfer. Today, the Aadhaar Payments Bridge that we discussed in a previous chapter, is being actively used to administer the LPG subsidy as well as other schemes across the country.

The time to carry out subsidy reform is right now. The domino effect of low global oil prices has reduced the government's subsidy bill and brought in a period of relative stability. More than two-thirds of the country already has an Aadhaar number, and the LPG subsidy system is proof that direct benefits transfers really do work. With low opposition to change and a smaller likelihood of increasing the consumer's economic burden, what better time can there be to initiate a sweeping overhaul of this millstone around the Indian economy's neck?

Capping cylinders, saving crores

Much of middle-class India relies on liquefied petroleum gas to cook its meals and, increasingly, to power its vehicles. Subsidized LPG is available through government-run Oil Marketing Companies (OMCs) via their national distribution networks, at an annual cost to the country of around $3.7B (Rs 240 billion). This fuel comes packaged in cylinders of different colours and sizes depending on whether they are meant for domestic or commercial use. Current usage statistics indicate that 120 million consumers (measured as families, not individuals) use an average of eight cylinders annually. Assuming that an average family consists of four people, that works out to a per capita consumption of two cylinders per person per year. As of 2011, 89 per cent of all LPG consumption was domestic, out of which the poorest 50 per cent consumed only 25 per cent of the available subsidized LPG.[24]

Until recently, an individual could buy an unlimited number of subsidized domestic LPG cylinders. As a result, people would buy these in large numbers and then sell them on the black market at the open market price, netting a handsome profit for themselves while bleeding the exchequer. Dealers created ghost consumers and delivery boys themselves false-booked cylinders under a consumer's name so that they could sell domestic cylinders for commercial use. The diagram in this section explains how the original LPG subsidy worked and outlines the many ways in which this system could be subverted.

In order to shore up the tottering LPG subsidy system, the task force recommended a multi-phase solution. The first step was the creation of a 'transparency portal' where every consumer's LPG consumption details were made available online, operating under the umbrella of the Right to Information Act.[25] The oil companies were initially hesitant to comply, but some pressure from the government ensured that such online portals were eventually created. Prominent politicians and industrialists were found to be consuming cylinders at improbably high rates—one politician was found to have claimed 240 cylinders in a year, or one cylinder every one and a half days, costing the central government almost $1,600 (Rs 100,000) in subsidies alone.[26] The media highlighted these stories and the resulting public outrage further fanned the flames for LPG subsidy reform.

The task force's second recommendation was to place a cap on the annual number of cylinders a customer could order. Initially set at six, the limit was later revised to nine. Customers were also requested to submit their Aadhaar details so that their identity could be verified. By linking the customer's Aadhaar number to their LPG connection, ghosts and fakes in the system could be eliminated; the cap on cylinders meant that customers would monitor their usage with a closer eye, further weeding out fraud. A study carried out by the International Institute for Sustainable Development reported that a cap of eight cylinders per year per household works out to an estimated savings of more than $615M (Rs 40 billion), or a sixth of the entire subsidy spend.[27]

The third recommendation of the Task Force was to set a single market price for LPG and transfer the subsidy amount directly into the beneficiary's bank account. In 2011–12, although India's three OMCs reported a combined turnover of $123B (Rs 8 trillion), their profits were only $1B (Rs 61.7B), or 0.7 per cent of their turnover—an unsustainable business model. Direct transfer of subsidy benefits can turn these loss-making OMCs into profitable institutions, since they no longer have to bear subsidy costs as losses in their balance sheets. Eventually, moving the subsidy out of the supply chain and directly to the consumer will pave the way for new firms to enter this huge market. Consumers will benefit from higher service quality, competitive pricing, and innovation; for example, new players might offer cheaper cylinders made of light composite materials rather than the iron cylinders currently in use.

The last recommendation the task force made was to convert LPG to a targeted subsidy. Currently, everyone is eligible for subsidized LPG, whether they need it or not. By limiting the subsidy only to those who cannot afford to pay the market price for LPG, the government can potentially save considerable funds that can be deployed for other purposes—witness the 'Give It Up' campaign launched by the current government, exhorting those who don't need the LPG subsidy to give it up for the welfare of the country. This type of fine-tuning is possible only in a direct transfer system.

The implementation of an Aadhaar-based direct transfer model for LPG subsidies is already underway; the initial rollout began in June 2013, covering eighteen districts in ten different states of India.[28] It was rolled out nationwide on 1 January 2015 and now covers 120 million customers. An estimated 33 million LPG connections were estimated to be fake or disconnected, and an additional 1 million customers voluntarily gave up the subsidy. The country's chief economic advisor Arvind Subramanian said that the savings in 2014–15 could be as much as $2B (Rs 127 billion), while in 2015–16, they were expected to be $1B (Rs 65 billion),[29] being lower largely due to lower oil prices.

Under this system, the customer pays the full market price for an LPG cylinder. As soon as the cylinder is delivered, the subsidy is transferred to the customer's bank account using the APB. The accompanying diagram explains the new LPG subsidy architecture in greater detail.

Those who do not require the subsidy simply do not enrol in the scheme. In the six-month period between September 2013 and February 2014, an estimated $62M (Rs 4 billion) was saved thanks to consumers who chose not to enrol. Consumption of domestic LPG, which traditionally grew by 6–8 per cent annually, showed a first-ever decrease of 18 per cent. Over 600,000 duplicate connections were detected; removing these duplicates would result in a projected

THE ISSUE OF LPG **SUBSIDY**

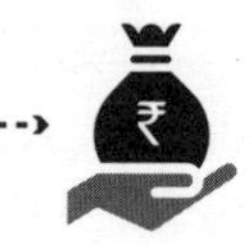

The government provides a direct subsidy on domestic LPG to producers.

The Oil Marketing Company (OMC) supplies LPG cylinders to distributors at subsidized rate.

The distributor allocates cylinders to customers at subsidized rate.

The customer buys LPG cylinder at subsidized rate.

The supplier/delivery person has an incentive for diversion.

THE PROBLEMS

PRODUCERS face a financial burden and focus on saving subsidy instead of quality

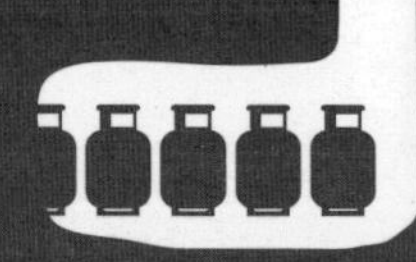

DISTRIBUTORS tend to divert subsidized goods for quick profit

CONSUMERS tend to overuse subsidized goods leading to shortages

Government refunds subsidy to deserving customer's bank account directly.

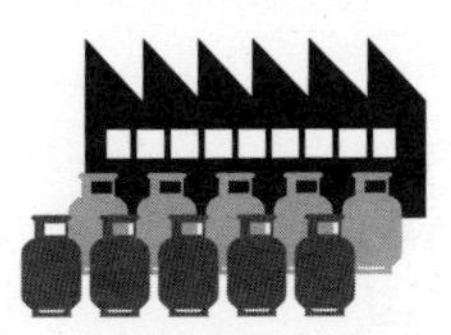

The OMC supplies LPG cylinders to distributors at market rate decided by demand and supply.

The distributor allocates cylinders to customers at market rate.

The customer buys LPG cylinder at market rate.

The supplier/delivery person has no incentive for diversion

THE IMPACT

Subsidy leakage reduced by 25%

Estimated savings of ₹6500 cr to ₹12700 cr in 2015

annual savings of $30M (Rs 1.93 billion). In total, the direct transfer system, in conjunction with the cylinder cap, was set to save the exchequer $1B (Rs 65 billion) a year.

Prior to the 2014 Lok Sabha elections, the scheme was suspended by the government, which appointed a committee to study its effects. The Dhande Committee studied the effectiveness of the direct benefits system, and strongly recommended that it be restarted, with some minor tweaks to make it more consumer-friendly.[30] Now branded as PaHaL (Pratyaksha Hastaantarit Laabh, or Direct Benefits Transfer), the direct transfer scheme was restarted and rolled out on an all-India basis on 1 January 2015. Prime Minister Narendra Modi commented, 'The PaHaL Yojana will bring an end to black marketing, and subsidies will reach people more effectively. Its role in nation-building is important.'[31] Thanks to these initiatives, for the first time, three hydrocarbons—petrol, diesel and LPG—are sold across India at market price.

Promoting agriculture, saving the environment

Viral visited the Gujarat State Fertilizers and Chemicals Ltd (GSFC), a Vadodara-based urea manufacturing plant, to get a first-hand picture of how the fertilizer subsidy worked on the ground. Unsurprisingly, the process turned out to be incredibly intricate and cumbersome. The government would sit down with fertilizer manufacturers to figure out the estimated demand for a given year, after which it created a 'movement plan', detailing how the fertilizer would be transported across the country, and provided freight trains to GSFC. Freight would be subsidized only if the company stuck to the official transport plan. Government-approved methods would be used to calculate the manufacturing cost, and the government would reimburse GSFC for the difference between this number and the subsidized retail price. The final sale price itself was determined by a piece of legislation that specified subsidies for ninety-three different grades of chemical fertilizer. The reimbursement process was complex and beset by delays, so every factory equipped itself with an army of accountants to chase down their money; an equal number of number crunchers in the government worked on tracking these payments.

Such an antediluvian system can hardly be expected to function smoothly. All across the country, farmers regularly complain of fertilizer shortages, particularly urea. Speaking to the owner of a fertilizer shop at a village near Vadodara, Viral was told, 'The demand is so high that as soon as the urea arrives, people buy it right off the truck—it doesn't even enter the warehouse.' Urea makes up nearly half of India's fertilizer market, and 80 per cent of all urea distributed and consumed is produced indigenously. The demand for urea is clearly driven by the fact that it's cheap—other, unsubsidized fertilizers like diammonium phosphate are in abundant supply, but farmers are reluctant to use them since they cost more. Dumping nitrogen-based urea on their fields, farmers ignore the fact that they need to include other types of fertilizer in their arsenal if the soil is to receive the correct mix of nutrients. Speaking about the overuse of nitrogen-based fertilizers relative to other nutrient types, Ashok Gulati of the Indian Council for Research on International Economic Relations says, 'This type of ratio is a disaster. It is keeping India from reaching the production levels that the hybrid seeds have the power to yield.'

The fertilizer industry is heavily dependent on worldwide market prices and the government's fiscal situation, a dependence that gives rise to uncertainty and unease. Progress is often stymied by the financial disincentives that subsidies unwittingly create. Take the case of urea, which can be manufactured using either naphtha (derived from petroleum), or through a newer and cheaper process based on natural gas. Thanks to the inadequate natural gas infrastructure, plants in southern India still use the older naphtha-based process. Since it is more expensive, manufacturers receive a higher subsidy, leaving them with little incentive to move towards alternative manufacturing processes or to push for natural gas connections.

Those manufacturers who do use natural gas lay first claim to all resources, skewing the availability for other industries. When the production of natural gas in the Krishna–Godavari basin dropped, supplies destined for urea manufacture remained unaffected whereas other industries had to make do with less or none at all. Trying to reconcile the fluctuations in natural gas prices versus the production costs of urea results in a constant state of financial flux, so much so that no new investment has been made in the fertilizer sector, either by government or by the industry.

Fertilizer subsidies have helped ensure food security by maintaining a steady baseline of agricultural productivity. However, we've seen that artificially low prices set in place a vicious cycle of fertilizer abuse, degradation of soil quality and declining crop yields. Trying to balance the needs of farmers against those of the manufacturers has left the government footing an enormously expensive subsidy bill—nearly $11B (Rs 730 billion) in the 2014–15 Budget—while the entire fertilizer industry stagnates in the absence of any incentives to improve production, and agricultural productivity deteriorates thanks to indiscriminate and unbalanced fertilizer use.[32]

Dwelling upon all these concerns, the Task Force on Direct Subsidy Transfer made a series of recommendations. As in the case of LPG, the task force recommended the creation and implementation of a fertilizer transparency portal which farmers can use to track the availability of stock. The window for the application of fertilizers is short; farmers sometimes have to queue for days to buy fertilizers, and still may not receive adequate supplies, as we saw in the case of Kripa Shankar. A transparency portal will eliminate long queues and waiting times, and allow for better planning so that crop health is unaffected. A second recommendation was the sale of fertilizers at market price, first to retailers, and later to farmers themselves, using a direct transfer model.

Food and fuel—the basic necessities

In 2006, Seethamma from Manjunathpur, a slum area in Mysore city, would get up at four in the morning with the hope of securing cooking fuel for her family. She walked with a fuel can to the public distribution shop (PDS) in Yadavagiri, which is two kilometres away and would wait for four to five hours until the kerosene cart arrived. She would be joined in the long line by many others, including schoolchildren. Even after such a long wait, she may not get her monthly quota of 6 litres of kerosene (reduced from 8 litres), as the kerosene cart may not come at all. She would follow the same routine in the following days until the kerosene cart arrived. At the time, PDS kerosene was sold at 14 cents (Rs 9) per litre, whereas diverted PDS kerosene was available in the black market for 54 cents (Rs 35) per litre.

Seethamma's story was recorded in a 2010 study carried out by the International Institute for Sustainable Development that examined India's attempts at kerosene subsidy reforms.[33] It is an excellent illustration of the many problems bedevilling the Indian government's efforts at providing cheap domestic fuel to those too poor to buy LPG. By supplying subsidized kerosene to people below the poverty line, the Government of India effectively insulates consumers from the volatility of world oil prices. For example, the kerosene subsidy per litre jumped from 26 cents (Rs 17) in 2011 to 43 cents (Rs 28) in 2012, thanks to an upsurge in world oil prices. The customer never noticed the difference since the subsidized price remained fixed; the resulting financial burden fell entirely upon the government. In 2013–14, the nation's kerosene subsidy bill was over $4.6B (Rs 300 billion).

While the cost of the kerosene subsidy is borne by the central government, the PDS that actually supplies kerosene to consumers is administered by the state. As subsidized kerosene wends its way from supplier to consumer through multiple levels of government, opportunities for pilferage and adulteration arise alongside. A landmark study by the National Council of Applied Economic Research (NCAER) found that 38 per cent of all subsidized kerosene ends up in the black market. The fiscal cost of the leakages within the PDS kerosene distribution network add up to an estimated $1.5B (Rs 100 billion), and in a further indictment of the system's inefficiency, cutting the PDS allocation by as much as 41 per cent would still be enough to meet the country's kerosene consumption.

The massive discrepancy between the market value and the

subsidized price provides a fertile breeding ground for the emergence of kerosene mafias, who exploit this pricing system for their own gain. In one attempt at thwarting these criminals, a blue dye was added to subsidized kerosene so that it could not be sold for commercial use. However, the incentive for arbitrage was so high that people actually set up processing plants to remove the dye so that you couldn't tell subsidized and regular kerosene apart. The scheme was a failure and eventually the dye-addition programme was halted. Further incentive for subversion arises from the fact that the kerosene subsidy is not uniform across the country; a person holding a Below Poverty Line (BPL) card in New Delhi is entitled to 20 litres of subsidized kerosene, whereas the same person in Bihar can only get 3 litres a month.

The supply of subsidized foodgrains through the PDS is equally inefficient; leakages in rice and wheat distribution cost nearly $3B (Rs 200 billion), while more than half of all wheat entering the PDS network is lost to pilferage. The public hunger for PDS reform was amply demonstrated in the state of Chhattisgarh, where Chief Minister Raman Singh won back-to-back elections on the strength of his initiatives to revamp the state's PDS network. In another attempt at reform, a pilot project using Aadhaar to verify the identity of PDS beneficiaries was launched in two districts of Andhra Pradesh in 2012. Once the scheme was implemented, households received their payments ten days faster on average. According to government reports, the introduction of Aadhaar authentication saw a savings of 10–12 per cent in terms of unclaimed provisions, and a state-wide rollout could save the state government $369M (Rs 24 billion). The Economic Survey of 2014-15 pointed out that the value of the savings generated by plugging the leaks in PDS distribution was eight times more than the entire cost of implementing the programme. Harpreet Singh, commissioner of the Andhra Pradesh civil supplies department, was quoted as saying, 'This system is helping us in exactly measuring the stock movements at the shop level on a daily basis. It makes the whole process so transparent that there is no scope for hiding any information at any level'.[34] The entire PDS distribution network of the Krishna district in Andhra Pradesh now uses Aadhaar-linked authentication; monthly savings run to the order of $1.2M (Rs 80 million), and A. Babu, the district collector who has driven the implementation of this scheme, says that the system is the 'best gift to an administrator like me'.[35]

The recommendations made by the task force in the case of PDS subsidy reform were along similar lines as other subsidies. The first was to introduce the direct transfer scheme for kerosene subsidy in two phases, given that it is administered at both the central and state government levels. In the first stage, state governments could transfer cash to beneficiaries; in the second stage, the payments would be transferred directly to the individual's bank account.

What other steps could be taken to overhaul the PDS network? One suggestion came from the government-organized task force for an IT strategy for PDS. Investigating the use of technology to build a modern, efficient PDS, the committee recommended setting up a professional company, the PDS Network (PDSN), which would manage everything from procuring goods to supplying them to consumers. The creation of the PDSN was announced in the 2012 Budget speech, but the idea has foundered in the absence of a champion to promote it. The Food Security Act, designed to provide subsidized foodgrains to nearly two-thirds of the country has been signed into law; the time is ripe to revive the concept of the PDSN, building a distribution system that delivers benefits to deserving citizens while curbing pilferage and fraud.

Bridging the gap: Using Aadhaar to deliver payments

No attempt at subsidy reform can be successful unless it tackles one of the most fundamental flaws in the system—the lack of a uniform and reliable method to identify beneficiaries. With Aadhaar in the picture, the government now had in its hands a powerful tool that would allow them to deliver with precision subsidies to only those who needed them. Aadhaar came with another benefit as well: it could allow the government to track the movement of payments through the system. It could even be used to create a database that linked all subsidies to all Aadhaar holders so that you could now see which subsidies a particular individual was claiming. This kind of cross-linked database offers enormous possibilities for both data mining and fraud detection. For example, fertilizer subsidy trends can be analysed to determine when and where fertilizer stocks should be moved around the country. A pension recipient claiming maternal benefits is a likely instance of fraud, and should be further investigated. None of these insights would be possible without building an Aadhaar-linked subsidy database.

How does the Aadhaar-based direct subsidy transfer system work? This is accomplished through an innovative application, the Aadhaar Payments Bridge that we have mentioned earlier. The APB is a centralized payments platform that allows the government to send a payment directly to an individual using only their Aadhaar number and a payment amount.

Email IDs allow you to send a message to a specific person over the internet, and mobile numbers allow you to call a specific person over the telecom network. In both cases, it doesn't matter if the person is in the same city or on another continent. In a similar vein, the role of Aadhaar in the payments platform is to act as a 'financial address'. It allows payments to be sent to a specific individual, irrespective of which bank they hold an account with. On one side of the platform are multiple government departments, funnelling in subsidy and welfare payments. On the other side are multiple banks, allowing customers to withdraw money. Both sides use Aadhaar as the identifier, either to send subsidies to the right person, or to ensure that the right person is claiming the money from their account. The entire system is now transparent, and every rupee can be followed from the time it enters the system as a government-issued payment to the time it ends up in the hands of the recipient.

The idea of using Aadhaar as a financial address, and its role in the eventual creation of the APB, all sprang from a presentation that Nandan made to the RBI in 2009; a scheme that is now benefiting millions of Indians was outlined on a single slide. The idea gained further traction with the creation of the task force to investigate the question of direct subsidy transfer. At the time the task force was active and the APB was being developed, the central government-owned LPG companies were already fairly computerized; the ministry of petroleum and natural gas, which handled the national LPG subsidy programme, was also especially keen to reduce the subsidy burden on the exchequer. The entire LPG subsidy ecosystem was thus uniquely poised for transformation, and became one of the first to adopt the APB to deliver subsidy payments directly to LPG consumers.

The only APB in the country so far was created by the NPCI, primarily for use by the government. Dilip Asbe explains how it works. The link between an Aadhaar number and a bank account is created in two layers. The first layer maps Aadhaar numbers to the corresponding bank, and this information is held in the APB. The second layer maps Aadhaar numbers to individual bank accounts, and this information is held by the bank. He explains, 'We designed this two-layered scheme for greater flexibility. This design also helps to compartmentalize a person's private information—there is no central database with everyone's bank account information and Aadhaar numbers in the same place.' If a person switches to a different bank in the future and wishes to use this new account to receive subsidy payments, all they have to do is to instruct the new bank to update their details in the APB, and payments can continue seamlessly.

Viral and Rajesh Bansal from the UIDAI worked closely with the NPCI team to implement the initial version of the APB. Within a few months, the team had procured the necessary technology and obtained regulatory clearance from the RBI, a record time in government and fast enough to rival the most nimble of private organizations. How was this speed achieved? The team met several times, bringing in relevant people from the UIDAI, the NPCI, banks and government ministries. Rajesh Bansal explains, 'We all had to agree on a set of common standards and working practices and take everyone's concerns into account so that we could develop a solution that worked for everyone.'

'The APB currently has over 600 participating banks—more than the number of banks offering ATM services or mobile payments in

India today,' he continues. 'The APB is one of the largest interoperable payments transfer systems in the country, and this scale has been achieved within two years of operations.' Today, over 200 million Aadhaar numbers are linked to bank accounts in the APB database. In 2014–15, 170 million transactions were performed through the APB, and over $923M (Rs 60 billion) transferred, with an average transaction size of $6M (Rs 365).

Welfare that works

How can direct cash transfers help to clean up welfare programmes? In the private sector, we're already seeing electronic payment models that are easy to use; thanks to technology, entry barriers are low while methods like data mining and fraud detection are used to ensure that all transactions are trustworthy. Technology can help to build a welfare system that is inclusive, allowing deserving individuals to participate while working behind the scenes to eliminate corruption and fraud.

The rules of eligibility, even for well-funded social safety nets, are complicated and cumbersome; a lack of clarity creates grey areas which must then be sorted out by bureaucrats. This provides plenty of opportunities for bribery and fraud, usually targeting the most vulnerable members of society. Technology can help us do away with these rules altogether. People can declare their own eligibility while signing up, and digital processes can be used to check whether these people are indeed deserving beneficiaries.

In the current subsidy system, geography controls eligibility. Consumers cannot obtain goods from any ration shop of their choice, or go to any health centre for medical services. In the accompanying diagram, we explain how an Aadhaar-linked subsidy model can solve the problem of portability, breaking the monopolies that currently exist. Consumers can now obtain goods from any vendor of their choice at market price, and receive the subsidy reimbursement directly into their bank accounts. This change will bring free-market dynamics into the formerly stultifying subsidy network, driving change in favour of the consumer.

The LPG subsidy programme makes a strong argument in favour of Direct Benefits Transfer (DBT)-based subsidy reform, with the projected savings from this scheme alone running into billions of rupees. The same model, when implemented for other subsidy programmes, can help to drastically cut India's subsidy burden. An additional push towards such a system has come from the recent Economic Survey, which has advanced the concept of the 'JAM Number Trinity'—a combination of the Jan Dhan Yojana, Aadhaar and Mobile numbers—as a solution for targeted subsidy delivery. The idea itself is not new; as far back as 2010, Nandan had stated, 'The slogan of *bijli, sadak, pani* is passé. Today it's about virtual things like UID number, bank account and mobile phone number.'[36]

Since that time, these three numbers have reached most of the Indian population. Today, 900 million Indians have an Aadhaar number, and this number is expected to rise to a billion by the end of 2015.[37] Over 200 million Aadhaar-linked bank accounts have been created. Finally, India has over 900 million mobile phone users. The combination of these three factors allows for the development of newer and more sophisticated mechanisms for the direct transfer of benefits, including mobile-based payment platforms or via India Post. Finance Minister Arun Jaitley referred to the concept as a 'game-changing reform',[38] and in the words of the Survey, 'If the JAM Number Trinity can be seamlessly linked, and all subsidies rolled into one or a few monthly transfers, real progress in terms of direct income support to the poor may finally be possible.'

One of the most important lessons that our experiences taught us was that the success of any reform effort depends on institutional capacity. LPG subsidy reforms were successful because they were rolled out by oil companies that were state-owned but boasted of professional management and computerization of the entire supply chain and distribution all the way to the last mile. By contrast, reform of the fertilizer and PDS subsidies will require intensive coordination between the central and state governments. The entire distribution chain is not yet computerized and there is no single database of all consumers eligible for the subsidy. This is essentially why cleaning

TAKING SERVICES NATIONWIDE

Most government systems lack choice

A ration card is only valid at one ration shop. This creates artificial monopolies.

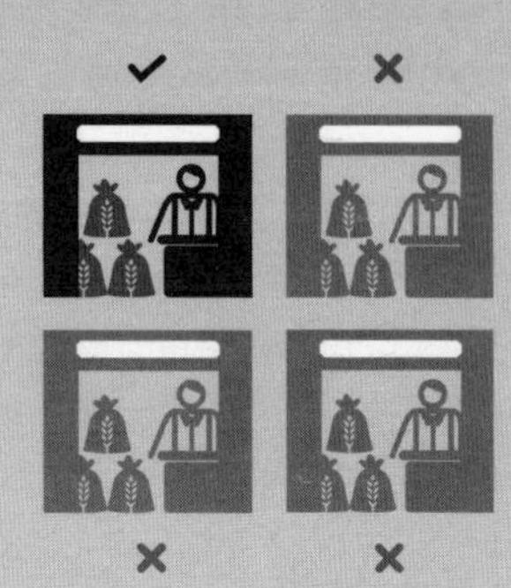

A Janani Suraksha Yojana (JSY) scheme payment is valid only at one Primary Health Care (PHC) centre.

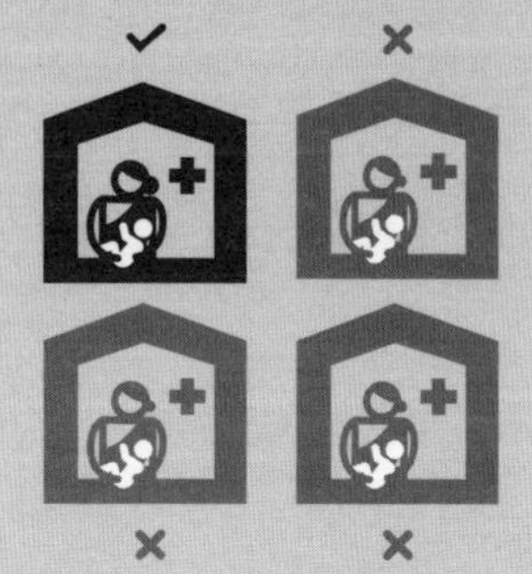

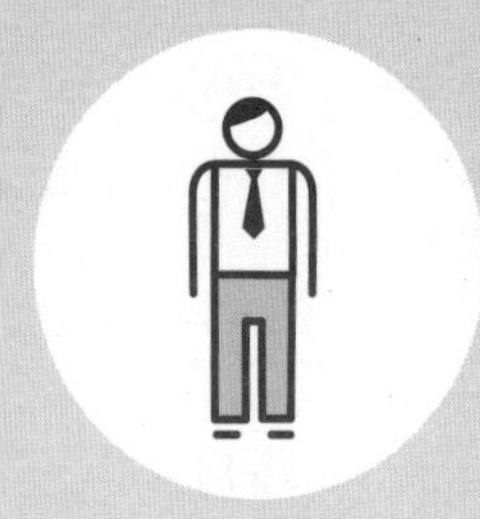

Migrant citizens cannot vote since their voter ID is tied to a single location, usually their hometown.

Bringing in choice through technology

With a portable Public Distribution System and ration card, a person can get affordable food from any ration shop of their choice.

With portability, a pregnant woman is able to access healthcare at any clinic. With electronic health records on the cloud, the clinic can easily access her information.

With a portable voter ID, citizens can vote from anywhere. Authentication at the time of voting eliminates fraud for free and fair elections.

up the fertilizer and PDS subsidies will be a far more onerous task.

Subsidies are inherently meant to address social imbalances, and are clearly well intentioned. However, failures of policy and implementation, coupled with outright corruption, have turned a well-meaning idea into an albatross around the government's neck, a burden which is impossible to carry and yet cannot be jettisoned for fear of public outcry. Technology-based solutions, implemented through well-designed institutional mechanisms, are a must for the widespread overhaul of our entire subsidy system, correcting for errors of exclusion and inclusion and eliminating fraud, while restoring the power of choice to where it has always belonged—in the hands of the consumer.

5

GOING COMPLETELY PAPERLESS WITH E-KYC

Paperwork is the religion of the civil service.

—Jonathan Lynn and Antony Jay, *Yes Minister*

NOT TOO LONG ago, opening a bank account in India meant spending a few hours at your local bank branch, clutching a file filled with photocopies of every single document you possessed that could prove who you were, when you were born and where you lived. Now consider an advertisement for ICICI Bank's new 'tab banking' service, launched in 2013. Bollywood superstar Amitabh Bachchan is flying kites with a gaggle of young men on a rooftop, when one of them gestures to a nattily dressed executive sitting at a distance. 'Who's that?' he asks. 'Oh, he's from ICICI Bank, and he's here so I can open a bank account.' 'Don't you need to go to the bank for that?' asks another of his companions. Bachchan, busy trying to down a kite, laughingly replies that if he had to go to the bank, he wouldn't have time for kite-flying. Taking a break from the action, Bachchan dashes off to meet the executive, who has already filled out the necessary form. He needs a photo, and whips out a tablet device to take a picture of his famous client. He then asks for a photo ID and address proof. Bachchan hands over his Aadhaar card, which is duly photographed. And that is the entire enrolment process. By the time Bachchan makes it back to the rooftop, he's already received an SMS thanking him for choosing to bank with ICICI.

Going from TV adverts to ground reality, consider a new facility launched by Axis Bank, which kicked off in Andhra Pradesh's Adilabad district in February 2014.[1] The customers here were not instantly recognizable movie icons but poor, weathered villagers, part of the nearly 40 per cent of rural India that continues to be excluded from the formal banking sector. They carried only one document with them—their Aadhaar card. (If they had memorized their number, they wouldn't even have needed that.) Their Aadhaar numbers and fingerprints were captured and authenticated with the UIDAI. If this information was correct, they would provide their consent for the UIDAI to share with the bank a digital copy of their photograph, name, address, date of birth and gender, all required for opening an account. A short while later, the customer would walk out with a transaction slip bearing the number of their brand-new Axis Bank account. A process which used to commonly take fifteen days and cost the bank anywhere between 77 cents (Rs 50) to $1.5 (Rs 100) for every new customer was now completed instantly, for a fraction of the original cost.

These are just two examples of the remarkable savings in time and money that are possible if our deeply entrenched dependence on paper-based documents is broken. Whether it's opening a bank account or buying an insurance policy, any service that requires identity and address proof should be just as simple and hassle-free as in the two cases above. In this chapter we discuss how Aadhaar-linked electronic Know Your Customer services are the first step on the path towards achieving these milestones.

Drowning in a sea of paper

In India, every stage of our lives is neatly bracketed by a document. When we are born, we get birth certificates. We graduate from school, and get school-leaving certificates. We pass major exams, and receive diplomas. We learn to drive, and get a driver's licence. We become eligible to vote, and register for voter IDs. We graduate from college, and get degree certificates. We get married, and get marriage certificates. If we buy a vehicle, we have registration certificates to prove our ownership. We insure our cars, and our lives as well, collecting insurance policy documents along the way. If we need to travel out of the country, we get passports. We obtain PAN cards so that we can pay income tax. We get ration cards; if our income is below the poverty line, we get BPL cards that make us eligible for government benefits. Some of us get caste certificates so that we can access the benefits due to us. We buy houses or land, and acquire a whole new bouquet of ownership documents, payment receipts and tax forms. We have children, and the entire cycle repeats itself. And when we die, it's a death certificate that marks the full stop to our lives.

Every stage we pass through requires documents from the previous one. You can't go to school without a birth certificate. You can't go to college without a school-leaving certificate, and you can't get a job without your college degree certificate. Even when you die, your relatives need your death certificate so that your assets can be managed correctly. Now imagine the plight of the millions who possess either no documents at all, or only one or two in the laundry list we've mentioned above.

While the government has yet to break its dependence on paper, in many other areas we have smoothly transitioned to the digital realm. A remarkable example is that of share certificates. Today, all of our share holdings are completely electronic, a switch completed nearly a decade ago. This was made possible by the Depositories Act (1996), which paved the way for the creation of the National Securities Depository Ltd. (NSDL), where digitized or dematerialized share certificates are now stored and accessed electronically. Share trading also moved from open outcry markets with paper trails to electronic trading. When an electronic trade is completed, the shares are automatically transferred from the seller to the buyer.

Similarly, our bank accounts have largely become paperless. We no longer write cheques or money orders but make payments digitally. We no longer have to receive bank statements by post. We no longer buy railway, bus or airline tickets in a paper format. We buy insurance policies online and the insurance certificates are often in electronic form. Today, many banks have started offering loans online with an electronic approvals process, possible because a person's financial history and credit score are accessible online for the bank to scrutinize. Most of our routine financial transactions can now be completed electronically without our having to physically step into a bank branch.

Breaking a bad habit: Building a paperless government

On the one hand, paper is the information currency of our government; the image of the 'sarkari' file stuffed to capacity with documents still endures. And yet in other parts of our lives, we have long since transcended the need for paper-based transactions. Instead of this uneven distribution, what if our entire society could be completely paperless? What if you never needed another piece of paper to interact with the government? What if all your information was stored in a secure but accessible electronic format, and what if this information could be combined in different ways to offer you innovative new services?

A digital identity

None of these visions can be realized unless all residents of India are in possession of a secure digital identity. This is where Aadhaar enters the picture. It is the only official document that exists in both paper and electronic forms. Most government-issued documents are in the form of an official card, which must be produced when required—the entire value of the document resides in that piece of paper. On the other hand, Aadhaar was deliberately conceived as a number with information linked to it, with no restrictions on what form it could take—it could be either paper or electronic, depending upon the needs of the individual. It could be printed on a government-issued card or made available online in a secure digital format. This was a very hard sell to many in the government, who were completely unfamiliar with the idea of a government document that could exist entirely in the digital world. As Ram Sewak Sharma memorably puts it, 'I tried giving the analogy that the cards were like the body and the number was like the *atman* (soul) and when the atman is available on the cloud, you do not require intelligence in the body!' As we shall see, the entire e-KYC system could be implemented only because Aadhaar can be used in a digital form; with a paper-based ID, e-KYC would have been impossible.

In order to prove your identity as part of the KYC process required for everything from opening a bank account to buying a mobile SIM card, you are asked to provide proof of your identity and proof of your address. It is these two documents that Aadhaar replaces. This may not seem like much, and certainly isn't the kind of radical reform that typically makes headlines. But add up those two pieces of paper across the millions of KYC transactions taking place every day in India, and the numbers begin to speak for themselves.

Cheaper services for all

Why is there an increasing interest in the idea of going paperless? For one, paperless products are cheap. The biggest expense is to develop and implement the technology that makes these products paperless, which is still much less than the cost of collecting, verifying, authenticating, storing and retrieving paper-based records. As costs come down and products get cheaper, more consumers can afford them, promoting inclusivity—look at the mobile connectivity boom in India spurred at least in part by cheap talktime recharge options.

A prime example of cutting both paper and costs is in the case of opening bank accounts. Every account that a bank opens for a customer comes with the cost of collecting documents to verify the customer's information, storing and retrieving this information when needed, and ensuring compliance with government regulations around the KYC process. Given these costs, it makes financial sense for a bank to open accounts only for those customers likely to maintain a relatively high balance. In practical terms, this means that banks have few incentives to offer services to the poor, who are likely to have lower balances in

TODAY	TOMORROW
All transactions are paper-based making millions of people invisible to the system.	With digital transactions, millions of people who had no presence now become visible to governments, banks, employees, lenders and more.

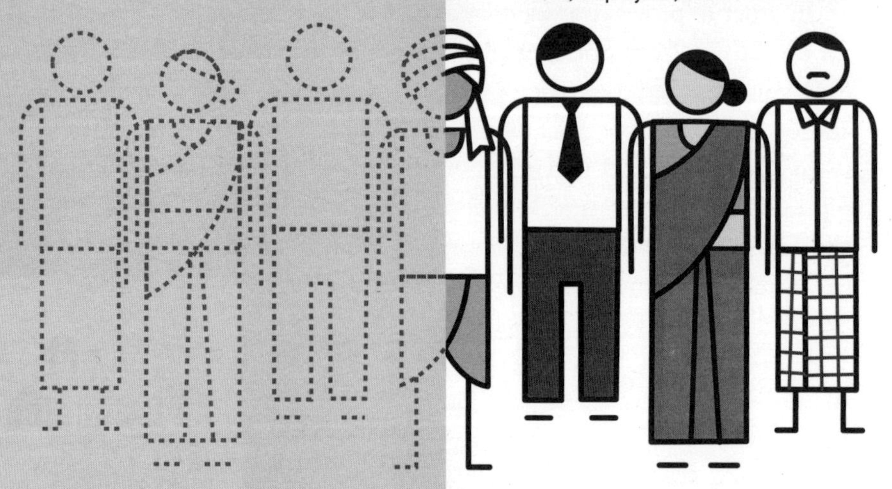

TODAY		TOMORROW
Remittances are sent as cash or as havala.	1	Remittances through the bank/post office become recorded transactions.
Transactions are heavily cash-based.	2	Cash use decreases over time, creating a cashless society.
Loans are taken from unorganized sources like the corner moneylender.	3	Loans and other financial products can be obtained from formal financial institutions.
No participation in the formal financial system, no record of transactions.	4	Financial inclusion increases, electronic payments reflected in banks create a transaction trail.
All commerce (farm produce sold, small enterprises) is in cash and unrecorded.	5	Formal tax returns filed via the GSTN make businesses visible, allowing them to get credit ratings and affordable interest rates on loans; tax evasion drops.

their accounts, assuming they even have one. In effect, the regulatory costs around KYC act as a barrier preventing financial services from reaching the poor and disadvantaged.

If the same process could be made entirely paperless and electronic, banks could now offer financial services to people like the villagers of Adilabad. Just as mobile phone companies came up with the idea of ten-rupee mobile top-ups to serve poor clients, so also banks and financial institutions can come up with innovative services to target this new class of customers. Digital information can be mined and the flow of money through the system can be easily tracked. Remittance records can be used to develop a credit rating system for all customers. Small businesses that formerly ran on cash will now have electronic tax payment records, making them credit-worthy. Millions of people who are invisible to the government and the system at large can now become visible in one fell swoop thanks to technology, as explained in the diagram on the facing page.

Buying biscuits and pension plans

As paperless KYC processes become the norm and financial services begin to spread across the population, what kinds of changes can we expect to see? Imagine being able to remove the friction in opening a small bank account, issuing a micro-insurance policy or selling a micro-pension product or a small investment product like a micro-mutual fund. Imagine that, instead of meeting a representative from a bank or insurance company, a customer can go to their local kirana store and complete an e-KYC process to buy any financial product he wants, from shares in a company to a pension plan.

This scenario isn't that much of a stretch when we consider the fact that millions of people use exactly the same operating model to top up their mobile phone balances. The only paper involved in the process is when the customer is buying a new SIM card; a one-time KYC process must be completed at this stage. Account statements and balance updates are all sent electronically to the customer via SMS, obviating the need for paper statements. Top-ups are done electronically at kirana

shops. Multiply a low-cost top-up by the millions of people carrying out such transactions, and we can see why the Indian market has been such an attractive target for telecom companies.

Today, your local kirana store also doubles as a mobile recharge centre; equip it with a microATM to carry out electronic customer verification, and the store now becomes a centre for basic banking transactions. People can deposit and withdraw money, pay bills, send remittances, pay their insurance premiums, contribute to their pensions or perhaps buy a slice of corporate India in the form of shares and mutual funds. As the accompanying diagram shows, customers of Priya Stores can now pay their bills and withdraw money with the same ease that they buy biscuits, toffees and fizzy drinks.

The authenticity guarantee

Apart from the ability to reach those segments of society who have been cut off from access to formal financial services, paperless products also make it easier to verify the authenticity of each transaction. Electronic documents come with in-built security measures that make them tamper-proof—for example, documents with a digital signature cannot be modified or forged in any way. With this level of safety, trust in the system increases and brings down the cost of doing business for everyone. If you're buying a product from a financial company today, regulators require that you be physically present while the transaction is carried out and while the company representative is verifying your documents. The e-KYC service does away with this requirement, since the entire transaction can now be completed online.

New connections, new ideas

The largest benefit of e-KYC comes from the fact that different forms of electronic data can now be combined to offer an entirely new class of services that could not exist otherwise. The potential impact is huge, especially in a society where people have to prove their identity, trustworthiness and reliability over and over to different institutions. For example, the computation of a credit score combines data from multiple sources, making it possible for those with a good credit score to get low-interest loans. Car insurance providers access your driving history to assess the level of risk and provide the most competitive pricing for an insurance policy.

Combine the speed of analysing digital data with the ease of verifying a digital identity, and you have a model in which transactions can be completed in real time—in minutes instead of days. With the upcoming data revolution, smartphones will keep getting cheaper while data costs go down, so everyone with a mobile connection can easily access the internet, laying the foundation for an entire suite of electronic financial services to be offered across the country.

If this vision is to become a reality, it requires a combination of regulation and innovation in the marketplace. Data privacy is a concern, and no agency should be granted unfettered access to a person's private information. On the other hand, it isn't possible to develop new services and products without regulated access to consumer data. The job of balancing these two imperatives falls to a regulatory authority; regulators like the Securities and Exchange Board of India, the Telecom Regulatory Authority of India (TRAI) and the RBI provide oversight to specific industries, ensuring that the consumer's interests are protected while also mandating a certain minimal amount of data sharing by all industry members so that things function smoothly. The same model can be implemented when it comes to building and offering paperless products and services.

The legislatory and regulatory framework needed for the emergence of a paperless society is largely in place in India. The Information Technology Act of 2000 lays down the criteria that electronic documents must fulfil in order to be treated on par with paper documents, which we will return to later in this chapter. The Electronic Delivery of Services Bill, introduced in the Lok Sabha in 2011, requires all public service delivery to be digital within a period of eight years. While the bill is yet to be passed, some Indian states have already implemented similar legislation. A further validation of the move towards a paperless government came on 15 August 2014, with the launch of the Digital India initiative; among other goals, It envisions the creation of a secure, Aadhaar-linked 'digital locker' for storing important documents electronically, as well as an eSign service that allows for secure digital signing of these documents. It is now in the early stages of implementation.[2]

When regulation is no longer an enabler

KYC norms were originally set in place to guard against money laundering and other fraudulent financial practices, but they took on a new, darker dimension after the terrorist attacks of 9/11. As Suyash Rai, a researcher at the National Institute of Public Finance and Policy, explains:

> On 9 September 2001, Mohammad Atta hijacked American Airlines Flight 11 and flew it into the North Tower of the World Trade Centre. Tracing flows of money led to the observation that a high ranking official within Pakistan's Inter-Services Intelligence (ISI) had allegedly ensured more than USD 100,000 was wired to Mohammad Atta, before the attack took place. Law enforcement authorities became quite keen to observe and block the 'financing of terror'.[3]

New laws against money laundering were implemented and existing laws were strengthened so that suspicious financial transactions could be closely monitored for terrorist links. The Financial Action Task Force (FATF), an inter-governmental organization set up to develop policies against money laundering and terrorist financing gained greater importance. India joined the FATF in 2002, which meant that it would have to comply with the international standards set in place by FATF. Such a decision made sense both in terms of security as well as economics—for example, Indian banks which were not FATF-compliant could not expand into other countries. As part of compliance with these regulations, India passed the Prevention of Money Laundering Act in 2005 and set up a Financial Intelligence Unit (FIU). Banks are supposed to report all suspicious transactions and customer behaviour to the FIU for further investigation.

As part of these heightened security measures, the RBI issued a directive that all new accounts opened in the second half of 2002 had to be compliant with KYC standards, and the same rule was extended to cover existing accounts by 2004. In practical terms, what this meant for customers was that they now needed to produce a valid proof of identity and proof of address. Without a passport, a PAN card, a driver's licence, a ration card or other such documents, it was no longer possible to open or operate a bank account with a recognized bank. At one stroke, the barriers for entry into the formal financial sector became impassably high for millions of Indians.

Even before the KYC norms were implemented, rural India had very limited access to banking facilities—some 600,000 villages had no access to banking services. The reason was largely economic—it is expensive to set up and staff a bank in a remote village, and the transaction volumes may be too small to cover such costs. Adhering to KYC norms is both costly and time-consuming, and this additional burden made banks even less enthusiastic about expanding into the rural sector. Suyash Rai continues, 'India is a member of FATF, and Indian regulators are obliged to apply Customer Due Diligence (CDD). Regulators in India have applied CDD through excessive forms of "Know Your Customer" requirements, which go well beyond the requirements of CDD. As a result, financial firms in India face increased costs.'

Usha Thorat, former executive director at the RBI, adds, 'Realizing that KYC was becoming a problem, the RBI mandated that small-value accounts—50,000 rupees or less—could be opened on behalf of an individual by a registered account holder who had already undergone the KYC process. Even though this was designed to make it easier for people to open accounts without getting caught up in KYC requirements, the banks were too worried about the FATF and the FIU to act on this decision.'

Another challenge came from the security concerns of the home ministry. Given the proliferation of prepaid SIM cards and cybercafes, anti-social elements could use phone networks and the internet for nefarious purposes without the fear of being traced, whether they were petty criminals or terrorists planning a major strike. To guard against this possibility, every customer buying a SIM card or using an internet connection must go through some form of identity verification.

The imposition of these stringent KYC standards has had the

unintended consequence of distorting the market for financial service providers. The people who most desperately need financial, telecom and internet services have found themselves out in the cold thanks to the prevailing regulatory climate, while the same regulations also make it financially unviable for providers to offer them these services. Our early discussions around the role of Aadhaar in KYC therefore centred around trying to solve both parts of the problem at once, providing a service that could successfully meet the competing needs of inclusion and security.

The journey towards electronic KYC

One of the earliest discussions we had about the need for electronic KYC was with Adhil Shetty, co-founder of BankBazaar.com, an online portal which helps banks disburse loans online. He told us that all parts of their loan disbursement process had been digitized, including getting customer details, salary details, bank details, the credit check, and even the bank's decision on the loan. There were only three parts that they had not been able to digitize due to regulatory issues: a photograph, proof of identity, and proof of address. These documents had to be collected from the customers at a huge cost by sending a representative to collect the documents, verifying them and then often going back and picking up new copies in case of an error the first time around. If these requirements could somehow be satisfied electronically, the entire loan disbursement process could go digital. Such early discussions were crucial in shaping the need for e-KYC, a vision which finally became a reality in 2013 after a great deal of work on the ground.

We worked with the central and state governments to accept Aadhaar as a proof of identity and address for accessing government services and benefits. Many state governments, in consultation with the UIDAI, issued official notifications putting Aadhaar on par with other forms of government ID. For any document to be accepted as a valid KYC document in the financial sector, it must satisfy the requirements of the FATF and comply with the provisions listed under the Prevention of Money Laundering Act (PMLA). Both of these fall under the jurisdiction of the department of revenue in the ministry of finance. Hence, our quest to add Aadhaar to the list of officially accepted KYC documents began here, and UIDAI patiently followed up for several months to get the necessary two lines added to the appropriate legal documents of the government. Once the government notification was in place, all the major regulatory bodies—the RBI, the SEBI and the Insurance Regulatory and Development Authority (IRDA)—followed suit, issuing their own notifications. Today, Aadhaar as a valid KYC document has been accepted by the telecom regulator, financial institutions, Indian Railways and all state governments.

The widespread acceptance of Aadhaar as a KYC document was half the battle won. Initially, we thought that the simplicity and standardization that Aadhaar could introduce would be enough to lower the costs of doing business for banks and service providers who were otherwise struggling with adherence to KYC norms. However, a meeting with the Financial Intelligence Unit (FIU)— the central agency tasked with monitoring financial transactions to detect activities such as money laundering—proved otherwise.

The UIDAI was represented at the meeting by a three-member team, consisting of Ashok Pal Singh, Rajesh Bansal and Viral. We presented the use of Aadhaar authentication as electronic KYC. In this transaction, the resident would provide his address, for example, and the service provider would enter it into the system, and cross-check it against the UIDAI database: Is this the address of person X? The database would merely answer Yes or No. The UIDAI would not actually share any information with the service provider at all; any questions would be met only with a Yes or No answer.

However, we learnt that this model would not meet the KYC norms. The main objection was that all the data would remain only with the UIDAI; the service provider would not have any customer data to enter into their records. Service providers usually need to have their customers' information, such as their photograph, on file for audits and other operational reasons. A second objection was that every time the UIDAI's database had to be queried, the service provider would have to manually enter the details, such as typing in a person's address and then asking, 'Is this the address of person X?' Manual data entry provides plenty of room for error, and this could unnecessarily complicate the KYC transaction. Apart from the FIU, other agencies like the department of telecommunications and the TRAI also raised similar concerns.

Given these issues, it was time for us at the UIDAI to go back to the drawing board, trying to come up with a solution that prioritized customer convenience while also satisfying regulatory requirements. The solution that emerged was the development of a paperless KYC system—the e-KYC service, as it came to be known. Similar to our earlier proposal, the first step remains the same, that of verifying an individual's identity using their Aadhaar number and biometric data. The second step is where things change. Instead of manually querying the database, the customer authorizes UIDAI to release their demographic information—their date of birth, address and gender—and their photograph to a bank or any other service provider, who can now retain this information in their records. Ashok Pal Singh recollects, 'When we were talking to the FIU about conventional paper-based Aadhaar, we asked them, "What is your wish list for KYC?" They indicated that they wanted something which was totally online, virtual and instant, as well as being foolproof and non-repudiable. All of these qualities were incorporated into e-KYC.'

While it seems simple enough in retrospect, it took us a long time to arrive at a workable solution. The original design of Aadhaar hinged on the fact that once a person's data entered the database during enrolment, it would stay there. No data would be allowed to leave, and only Yes/No responses would be permitted during authentication and verification. The Aadhaar holder was the only person who could look up their own data and update it as needed.

The idea of any data leaving the Aadhaar database made many of us very uneasy, ingrained as it was in our minds that data could enter, but could never leave. It took us all a while to come to terms with the realization that the e-KYC solution was compliant with Aadhaar's security and privacy framework. This mental journey was

BEFORE — K.Y.C PROCESS — AFTER

BEFORE

1. Photocopy ID proof document

2. Photocopy address proof document

3. Self-attest copies

4. Originals verified at bank/service point

5. Banks/telcos create digital records and verify again

6. Send someone to verify address and other details

AFTER

1. Enter Aadhaar number and provide fingerprint/one-time password (OTP)

2. Request sent to Aadhaar server. If fingerprint/OTP match, server returns person's details.

3. Data verification is instant, for immediate activation

Long verification process

Verification cost is high (Rs 50–200)

Verification in a few seconds

Low verification cost. Can reach the poor

quite instructive, helping our thought processes to mature to the point where we could envision Aadhaar being applied in many different domains. As closely as the trio of Ashok Pal Singh, Rajesh Bansal and Viral had been involved with the KYC story, it isn't surprising that they were among the first to get comfortable with the e-KYC concept. Others were more resistant; Nandan himself was not fully convinced, and had asked our team to come up with an alternative solution. We were quite confident with the model we had proposed, and to move towards a final decision, we resorted to the time-honoured government tradition of 'putting it on file'. After the dust had settled, Viral recalls Ashok Pal Singh saying, 'Whoever does not like this solution can reject it on file.' Rajesh Bansal was also given to proclaiming, 'Nobody will ever reject a pukka file.'

This 'pukka' file, containing the e-KYC design and all correspondence with the ministry of finance, the FIU and various regulators, was by now thick enough to give an encyclopaedia a run for its money. The file turned out to be the catalyst; all the key people in the decision-making chain signed off on the design, budgets were allocated immediately and work on the implementation of e-KYC began in earnest. Say what you will about the government's archaic file system, there is a lot of power contained in those green sheets, thanks to the merits of clear documentation, traceable decision-making and long term record-keeping.

Launching e-KYC on the ground

To get a sense of the value of Aadhaar-based e-KYC in the field, we spoke with Gautam Bhardwaj, the founder of Invest India Micro Pension Services (IIMPS), a social enterprise that enables poor people to set up low-cost savings for retirement. According to him:

> In early September 2014, we launched paperless and cashless enrolments for pension schemes at Tumkur district in Karnataka. Roughly 500 low-income women members of self-help groups living in remote villages in Tumkur participated in two-hour

> retirement literacy group meetings over ten days. Of these, around 300 women have decided to join the micro-pension programme. Application forms for both products are digitally filled using data from each person's e-KYC. Each form is digitally signed through a biometric authentication. Contributions are loaded onto bank-issued prepaid cards and moved into an escrow account in real time. In all this, the field experience with e-KYC is simply wonderful! It takes 3.5 minutes on average to complete the full enrolment and payment process. Forms and contributions are then electronically transferred for processing.

In fact, IIMPS found the paperless KYC and payments solution so compelling that they are pushing for wider adoption despite connectivity issues. Says Bhardwaj, 'In some villages, we're facing an implementation challenge on account of internet connectivity although we've tried internet data cards of all telecom providers. We're trying to figure out some solution including signal boosters.'

Once the e-KYC system was created, we had to go back and make the rounds of government once again for it to receive official approval. As before, we started with the department of revenue, who themselves asked the ministry of law and justice to weigh in on whether the e-KYC solution could be considered equivalent to existing paper-based methods. Having received the legal go-ahead, the department of revenue issued an official notice accepting the use of e-KYC. The three major regulators followed suit, and banks, insurance companies and security firms started to accept the e-KYC system. Ashok Pal Singh points out, 'The collaborative approach that resulted in the birth of e-KYC is quite rare in the government. Instead of taking a hardened position and making it a turf war, the UIDAI worked together with the department of revenue and the FIU to come up with a KYC solution that met everyone's requirements.'

A major push for large-scale adoption of the e-KYC service came with the launch of the Pradhan Mantri Jan Dhan Yojana (PMJDY). Under the provisions of this scheme, e-KYC would be used at scale for the rapid opening of bank accounts. Banks are now insisting on the use of Aadhaar and e-KYC before disbursing the $77 (Rs 5000) loan that has been sanctioned by the PMJDY so that one person cannot take out the same loan multiple times across different banks. Today, the UIDAI handles over 50,000 e-KYC requests daily, and that number is projected to increase exponentially in the future.

The gift that keeps on giving

Let's take a quick look at why the e-KYC system that we at UIDAI came up with is a win-win proposition for both service providers and customers.[4] Cutting out paper cuts both costs and time; the service provider doesn't have to manage the documents of thousands of customers, and since all records are provided instantaneously in electronic form, the time taken for document verification and data entry is reduced to zero.

As a result, Axis Bank and the like can now expand into rural and underserved areas, a proposition which would have made little financial sense before the advent of the e-KYC system. The positive impact on financial inclusion, bringing in ever more people into the formal financial sector, is an illustration of one of our core ideas—expecting organizations to participate in an initiative solely because it's a worthy social goal isn't likely to succeed. The minute we manage to make it an attractive business model by adding appropriate financial incentives, people are immediately willing to join in, and the whole ecosystem grows and expands in ways we ourselves might not have foreseen.

e-KYC also represents significant improvements in information security and handling, especially important in a country that doesn't have a strong set of regulations around data privacy. Features like explicit consent, biometric verification and digital signatures make the e-KYC process robust and tamper-proof, and resistant to identity theft.

Transactions are easy to store and trace, making audits far simpler. And finally, the e-KYC platform provided by Aadhaar is open to all service providers, both government and private, to use in any transaction that requires identity verification, whether it's in obtaining an LPG connection or determining eligibility for a bank loan. e-KYC is also designed to answer the needs of a highly mobile society. Digitizing the KYC process provides an easy method of identity verification irrespective of a person's physical location.

For those of us who already possess two or three forms of ID and have easy access to banking facilities, the use of Aadhaar as a KYC document might make life a little simpler. But for the majority of Indians without bank accounts, Aadhaar-based e-KYC enables them to enter the formal financial sector on the strength of a number and a fingerprint. It is a validation of our efforts that the PMJDY, aimed at eradicating India's 'financial untouchability',[5] accepts Aadhaar numbers as being sufficient proof for KYC. By providing an open, Aadhaar-based KYC platform, we hope to revolutionize the way customers and businesses interact, whether in the public or private sector.

Opening up the formal financial sector to all Indian citizens is merely the first transformation that the e-KYC process can bring about. Ultimately, the power of e-KYC lies in the fact that it can be used for any conceivable application requiring an identity, an address, or both, to be verified. Cross-checking your biometrics might be all it takes for you to enter an airport or board a train. The same data could also be enough for you to buy a SIM card, operate a vehicle or cast your vote. We can expect a whole host of new business models to spring up around the e-KYC ecosystem, using the idea of an electronic identity to develop services and products that are also completely digital. e-KYC is the first step towards building a digital identity for every Indian, and to move towards the vision of India as a completely paperless society.

6

INTEGRATING OUR ECONOMY WITH THE GOODS AND SERVICES TAX

Taxes are what we pay for civilized society.

—Oliver Wendell Holmes, 1927

IT'S A MISTY WINTER morning in Khanauri, a state border checkpoint in Punjab, and Lakshman Singh is getting restless. The thirty-eight-year-old is driving a truck laden with chemicals, and has been waiting here for six hours to pass through the checkpoint. Why the delay? Singh complains, 'They say they have not received the information from the previous check-post about the passage of my truck. That's not my fault.' Nearby, Dinesh Kumar, who is ferrying a load of yarn from Gujarat to a blanket factory in Punjab, has to fend off anxious calls from factory managers asking where he has reached, and whether he is stuck. Around Singh and Kumar, drivers stand in long lines, waiting to receive an official go-ahead while their assistants huddle around fires and brew up tea to stay warm and stave off boredom.[1]

A similar story unfolds at the Walayar checkpoint, on the border between Kerala and Tamil Nadu.[2] Trucks routinely get stuck here for four to five days awaiting clearance; textile entrepreneur D. Bala Sundaram, who runs a business with an annual turnover of Rs 9 billion, is so fed up of the delays that he no longer sends his trucks to the nearby international container terminal in Cochin. Instead, it's more efficient for him to send his trucks hundreds of kilometres in the opposite direction to the smaller port of Tuticorin, from where the cargo will head to Colombo, Sri Lanka, and thence around the world. Even though this diversion drives up his freight costs by 20 per cent, Balaji says the trade-off is worth it. 'We can give our clients an exact date when the cargo will reach its destination. With the Cochin port, we simply cannot be sure when or whether our cargo will reach the port itself.'

India's roads account for two-thirds of its freight traffic, but only 40 per cent of travel time is spent actually driving. The rest is lost to waiting at border checkpoints, paying taxes—there are eleven different categories that apply to road transport alone—and dealing with the local authorities.[3] According to the World Bank, these delays mean that the logistical costs of manufacturing in India are two to three times higher than international benchmarks; cutting wait times by half would decrease the logistics costs by 30–40 per cent, making India's manufacturing sector more competitive globally.[4]

It's not just the manufacturing sector that has to contend with the complexity of India's tax laws—it applies to all taxpayers in the country. When it comes to taxation, for example, the relationship between the people and the tax authorities assumes an adversarial tone, extending even to government-produced adverts on the radio warning you that for every high-value purchase or investment you make, the taxman is watching. During his election campaign, Prime Minister Narendra Modi used the phrase 'tax terrorism' to describe the relationship between tax officials and the public, saying, 'You cannot treat every citizen like a thief. Some serious thought is required and financial experts must see how taxes can be simplified.'[5] The finance ministry was also expected to issue a directive to the tax authorities to avoid 'harassment of taxpayers, whether they are individuals or companies'.[6]

In the view of Bharat Goenka, co-founder of Tally Solutions, a provider of business accounting software used by hundreds of thousands of businesses and small merchants, and someone who is deeply engaged with India's taxation systems, 'The government always works on the principle that their job is to make the law, and the citizen's job is to follow the law. The process of ensuring that the citizen is capable of following the law is not anyone's problem. Following the law comes through the concept of enforcement rather than the citizen's convenience.'

Taxes are the price we pay for living in a civilized society—they pay for our military and police forces to keep us safe, they pay for our roads, our schools, electricity, water—in effect, every public good that our nation provides its citizens. But when they are not implemented well, they hurt both our economy and our people. We need a taxation system that has a set of transparent, clearly understood rules, making compliance easy and evasion difficult. In our opinion, it is technology-based reform that will deliver on these twin goals.

India's skewed markets: The United States of India

By any measure, the town of Baddi in Himachal Pradesh ranks as an economic success story; boasting of over 2000 factories, Baddi generates an annual turnover of $9B (Rs 600 billion), acting as an economic engine for the entire state.[7] And yet, by the rules of the open market, Baddi is a terrible candidate for a manufacturing hub—there is no rail transport to the nearest major city, Chandigarh, and local truck unions have enforced a monopoly that makes road transport an expensive proposition.

Then how did Baddi manage to become a manufacturing powerhouse? The answer lies in the fact that the centre doled out massive tax exemptions to factories set up there, including five years of exemption from income tax and ten years of exemption from central excise taxes. These 'tax holidays' have now come to an end, leaving the authorities worried. Newer companies have little incentive to adopt Baddi as a base, and existing companies might choose to scale down operations, resulting in a massive hit to the economy of both the town and the state. When the basic rules of setting up a manufacturing plant, such as easy access to transportation, are bypassed in favour of tax considerations, it is an indication that tax policies are not in sync with market forces, producing gross distortions as a result.

A combination of political and economic factors have seen India's manufacturing sector stuck in a long-term holding pattern, its contribution to India's GDP hovering around the 15 per cent mark for decades; that number is over 20 per cent for countries like China, Indonesia and South Korea which have outstripped us to become global manufacturing hubs. Part of the problem is India's skewed markets; Nandan described them in *Imagining India* as 'splintered' and 'fragmented', a legacy of blinkered economic policy decisions that continue to hobble growth and development.

The current state of Indian markets can perhaps be described as 'quasi-open'; capital and labour move freely across the country, while goods and some services do not. Services like banking and telecommunications already benefit from a common market; they can be set up anywhere in the country and sell their products to people without difficulty. On the other hand, our manufacturing sector is heavily driven by the local tax structure and tax breaks, which take priority over such factors as labour, transport and economics—an unsound and economically inefficient way in which to operate.

The idea of combining India's splintered and fragmented markets into a single unified common market is not new. Simply put, a common market allows goods and services to be manufactured where it makes economic sense, for them to be sold where the market demands, and for free, unrestricted movement of these goods and services across the entire country. Tax regimes should support this vision, with a simplified taxation system that abolishes the traffic-choked checkpoints at state borders. In a common market with a well-designed tax regime, tax considerations will no longer distort business decisions.

Tax reform is only one step towards building a common market. It must be accompanied by reform in several other sectors as well: farmers should be able to sell their produce freely in any market in the country; land reform and better land titling systems should create a countrywide market for land; and a common national grid should allow for easy transfer of surplus electricity between producers and consumers so that all parts of the country can have a robust electricity infrastructure capable of supporting manufacturing units. In this chapter, we focus on the issue of tax reforms, specifically the Goods and Services Tax; in the next chapter, we will examine the potential of an electronic toll system to allow for the smooth flow of goods nationwide. These reforms are essential if India is to overcome its abysmal performance in the manufacturing sector so far, and rise up to assume its rightful place in the global market.

Cutting the Gordian knot of taxation

A major step in untangling India's tax laws came with the introduction of the Value Added Tax (VAT) in 2005. But, as Bharat Goenka tells us, 'Simply announcing a government scheme without corresponding education on the ground will not lead to enthusiastic adoption.' And so his company spent 'an obscene amount of money promoting VAT', conducting workshops and melas all over the country for small businessmen explaining what VAT was and how it worked. Goenka says, 'We didn't sell a lot of our software but we ended up meeting almost 300,000 merchants across the country. Our workshops were so popular that even the tax authorities started participating.' Within a year of VAT being rolled out, he recalls, 'There was a palpable shift towards adopting VAT. Earlier, there was no linkage between purchases and sales—VAT created that linkage. The shift we sensed in our own customer interactions was not because VAT made it more convenient for people to pay their taxes, but because the linkage made it harder for people to avoid paying tax.'

Prior to VAT, what we had was a cascading tax system—every time goods were sold through the manufacturing chain, a sales tax was levied on the total value. As the accompanying diagram illustrates, taxes added up at each stage, and attempts to avoid this ballooning tax bill created distorted incentives for manufacturers. T. Koshy, a former executive director at the National Securities Depository Limited, who has worked extensively in the field of taxation, gives us an example. 'The cascading tax system incentivizes vertical integration so that the

VAT: A BETTER WAY TO COLLECT TAX

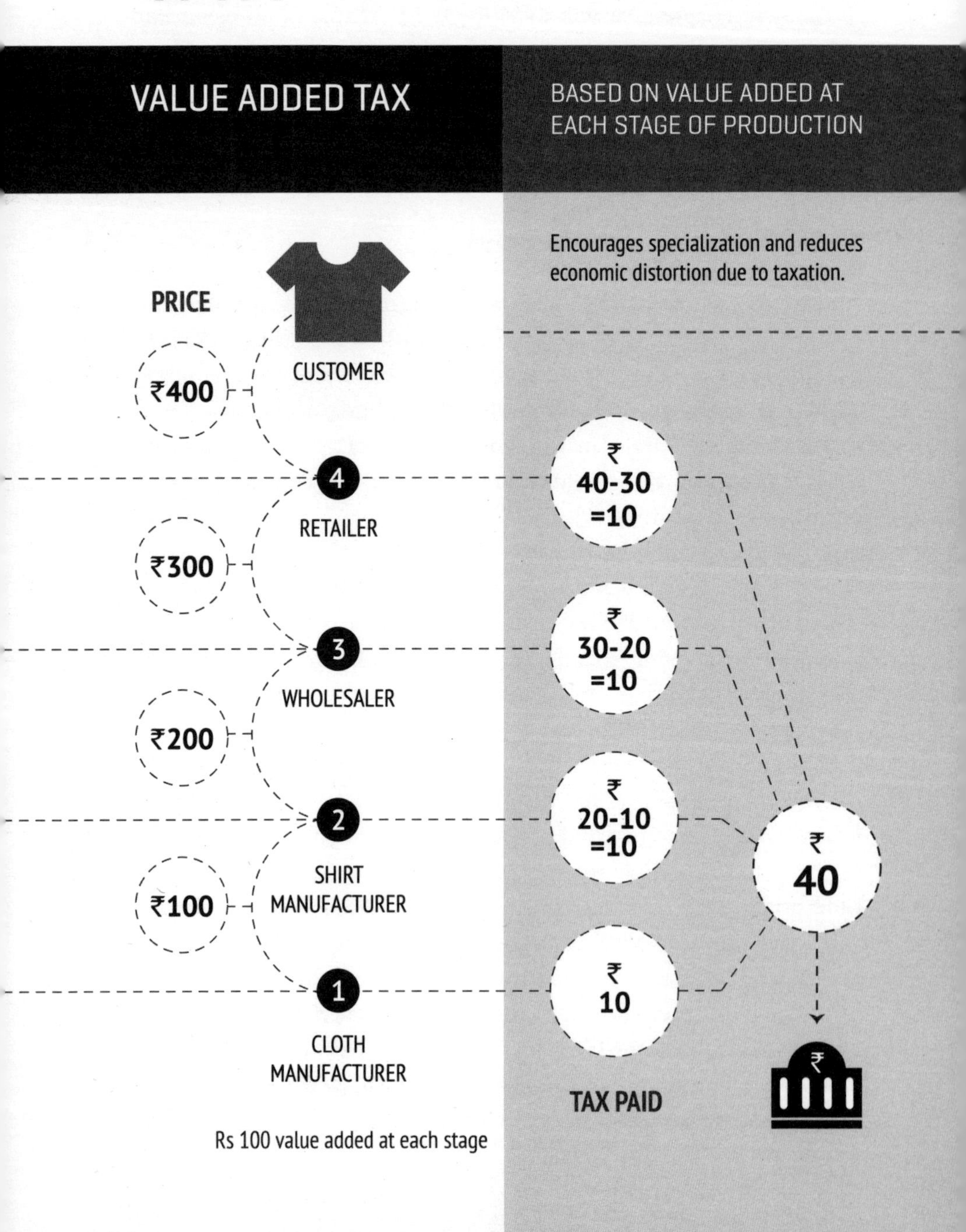

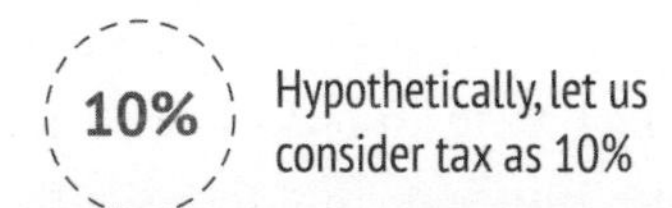

BASED ON SELLING PRICE, COMPOUNDING AT EACH STAGE OF PRODUCTION

CASCADING TAX

Incentivizes vertical integration to save on tax.

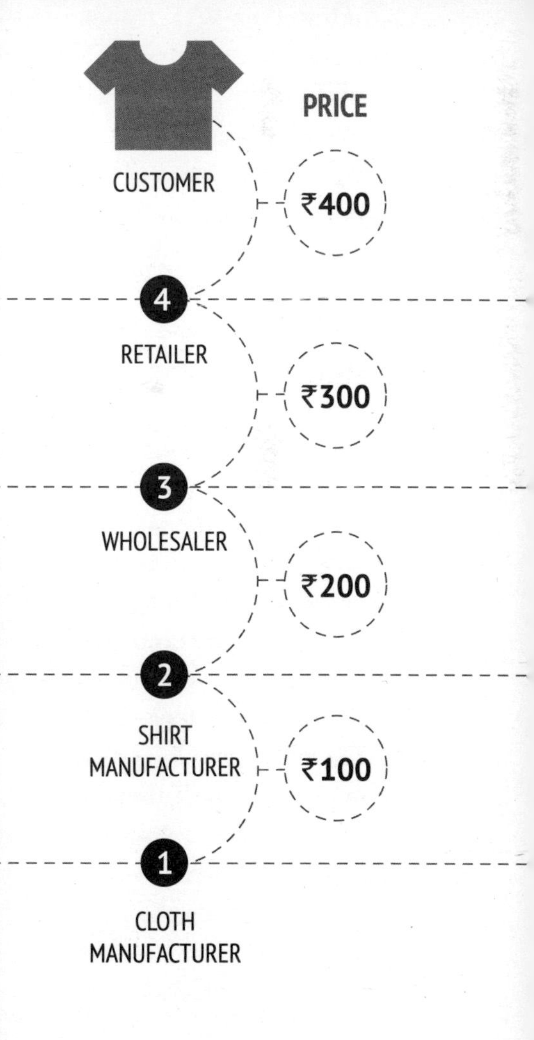

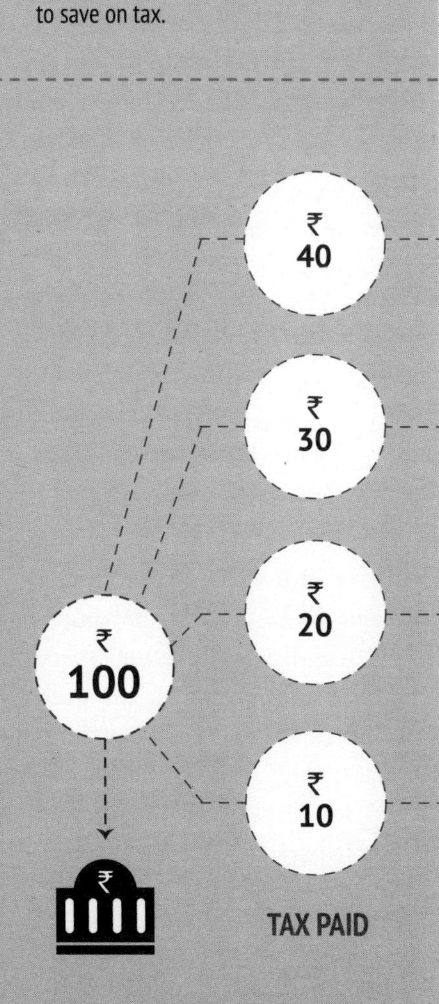

same firm tries to carry out as many of the intermediate manufacturing steps as possible in order to pay less tax.' In our example, that would mean one firm controlling the entire manufacturing and supply chain, all the way from making yarn to selling the finished shirt in a retail store. Koshy adds, 'In the cascading system, one item can have two different prices depending on whether it was manufactured by one large firm or many small firms. This price difference is solely due to taxation, and this sort of distortion is what the VAT system was designed to eliminate.'

How does VAT manage to eliminate these distortions? Just like the cascading tax system, VAT too is applicable at every stage of the manufacturing process. The critical difference is that it is not applied to the total value of the goods; rather, it is applied only to the value added at each stage of production. Equally important, the system is set up to be self-policing; the taxes from every stage feed into the next, and if one link in the chain fails—one manufacturer does not file taxes correctly, for example—the entire chain collapses. Since it applies only to the value added by each manufacturer and not to the total value, VAT also encourages specialization, rather than the kind of vertical integration that arose in the earlier cascading tax regime. Increasing the levels of transparency and making it harder to game the system result in increased revenue collection and a lesser burden of oversight upon the tax authorities. Koshy adds, 'VAT is a truer measure of economic growth than creating artificial incentives through tax holidays and exemptions.'

The creation and implementation of the VAT system was a Herculean task; as Nandan wrote in *Imagining India*, it was 'a convoluted and contested reform that started in 1997 and was negotiated every inch of the way. It involved two central governments, all the state governments with their many, divided loyalties, several enterprising bureaucrats and a couple of holdout states who resisted the reform well into 2007. All in all, it is a tale of a few good men negotiating and manoeuvring the reform for the better part of a decade, through the labyrinth of the government and its bureaucracy.'

The idea of VAT met with strong resistance from state-level

governments who feared the centre would now appropriate their revenue and tax collection powers, leaving them beholden to the centre and unable to function independently. Every state had to be convinced and brought on board before this new taxation system could be implemented. P. Chidambaram, the finance minister at the time, offered to compensate states for any losses incurred once the VAT system was in place. He assured the states that if they achieved anything less than 17.5 per cent in tax growth, the Government of India would make up the difference, creating a safety net for states who were otherwise reluctant to sign up for VAT implementation.

Unifying taxes, unifying markets

While the new system was an excellent start, it didn't address all of the many problems bedevilling India's tax regime. First, VAT operates only at the state level, which leaves untouched the problem of tax fragmentation between the centre and the state—the centre taxing goods at the point of manufacture, and the state taxing them at the point of sale. Moreover, this tax doesn't apply to services, which are taxed by the centre in the shape of service tax. There's still an unruly thicket of taxes to be navigated—the octroi system, central- and state-level taxes, entry tax, stamp duties, telecom licence fees, turnover taxes, taxes on electricity, taxes on goods and passenger transport, state-level cesses and more.

It's not that tax rates in India are particularly high—data from the World Bank pegged India's tax collection (as a percentage of GDP) at 10.8 per cent in 2012, compared to 10.2 per cent for the United States, and nearly 25 per cent for the UK and France.[8] The problem is that there are too many different types and rates of tax. Operating across various levels of government, such a multilayered taxation system leaves plenty of room for grey areas to arise. These uncertainties have to be resolved by the tax authorities, and as with any complex system that relies upon human judgement, a uniform implementation of the law is virtually impossible. Arbitrating tax disputes is also a waste of precious time and resources. It becomes harder to enforce compliance with the law, and a large bureaucratic machinery has sprung up to police taxpayers.

High taxes on import, as well as excise and other taxes on domestically produced goods, result in a lopsided structure of production, consumption and export. The flawed priorities that are the offshoot of our current tax system drive firms to lobby for favourable modifications in the tax schedule, making tax policies even more market-unfriendly. As a result, the pricing of goods is not an accurate reflection of the prevailing market forces, a worrying factor in a country where consumers are known to be remarkably sensitive to small fluctuations in price. The ultimate goal of tax reforms is to empower consumers to make rational choices which are based upon their own needs and preferences, rather than being the end-product of policy decisions taken by the ministry of finance.

Given these long-standing issues, the idea of a Goods and Services Tax (GST)—a single, indirect tax system encompassing both goods and services—has been doing the rounds from as far back as 2000, when Atal Bihari Vajpayee's government was in office. A task force report from 2003 laid down the original design ideas underlying the GST.[9] GST replaces the state-wise VAT, central sales tax, service tax and a constellation of other local taxes with a common taxation framework. Under this regime, there is no differentiation between taxation of goods and services.

The GST's design philosophy incorporates several other principles designed to, as Bharat Goenka puts it, 'create a system that solves the government's problems as well as the citizen's—the only way you can both create a law and ensure adherence to that law.' Apart from streamlining our tax system, the GST also possesses a second, critical attribute—it applies to goods at their point of sale, not at the point where they are produced, making it a 'destination-based tax'. In practical terms, this removes any incentive for states to offer manufacturing sops. Instead, they must focus on creating the kind of economic environment that spurs genuine growth and consumption. In a single stroke, this shift will give rise to a system in which economic efficiency drives the choice of location where goods and services are produced, rather than artificial incentives giving rise to tax havens like Baddi and its ilk.

Breaking the logjam with the Goods and Services Tax

How do you find out whether a government is serious about implementing a particular reform initiative? In our experience, the answer is to pay close attention to the speeches that its leaders make, whether it's the Budget speech in Parliament or the Independence Day speech delivered by the prime minister from the ramparts of Delhi's Red Fort. A mention in one of these key speeches is usually enough to signal seriousness of intent to the offices of the prime minister and finance minister, who then begin to keep track of progress made on these specific promises. While this is no guarantee that things will get done, an official, and highly public, endorsement from the country's leaders at least gets things moving in the right direction.

And so it was that GST implementation began to move forward with a mention in Finance Minister P. Chidambaram's Budget speech of 2005–06.[10] One big obstacle the GST had to immediately confront was the lack of a legislative framework to support it; if the GST were to become a reality, a constitutional amendment would need to be passed, allowing both central and state governments to concurrently levy taxes on both goods and services, while interstate commerce would be taxed by the centre. Constitutional amendments are difficult by design; both regional and national political support is essential, and while Aadhaar has shown us that consensus can be achieved across party lines, it is usually a long-drawn-out affair, moving in fits and starts.

While political consensus around the idea of the GST was being painstakingly built, the UPA government also decide to create a detailed plan for its implementation, updating the 2003 version. They needed to look no further than the model the previous government had used to drive the VAT reforms in 1999. Finance Minister Yashwant Sinha had created an empowered committee of state finance ministers, chaired by Asim Dasgupta, the finance minister of West Bengal. This was a move that Nandan described as 'a masterstroke in reducing dissent'; rather than issuing diktats from on high in the centre, the states themselves were put in charge of VAT implementation, speeding up rollout considerably. The UPA government followed exactly the same strategy for GST, down to appointing Asim Dasgupta as chairperson of the committee. In parallel, Finance Minister Pranab Mukherjee also announced the creation of a Technology Advisory Group for Unique Projects, chaired by Nandan. GST was one of the five key technology projects for which this group was to prepare a detailed blueprint.[11]

We both attended some meetings of the empowered committee, and got a ringside view into the delicate centre–state manoeuvrings around the design and execution of a new project in government. As head of the committee, Asim Dasgupta, the then finance minister of West Bengal, deftly directed the overall proceedings while taking on board all points of view, from the finance ministers of every state to finance secretaries and revenue officials.

The meetings took place in a conference room furnished with a large square table, and arrayed on every side were about ten finance ministers, beside or behind whom sat their state officials. For every point on the agenda, all state representatives were invited to share their opinion. We found states to be profoundly uncomfortable with the idea of the GST for two main reasons: they feared a loss in their power to levy taxes, and an overall loss in revenue if they moved towards the destination-based tax system that was proposed under the GST, especially for states heavy on manufacturing. For example, the state of Gujarat demanded that they be allowed to levy an additional 1 per cent tax over and above the GST; the state finance minister, Saurabhbhai Patel, said in a 2015 pre-Budget meeting, 'As we move closer towards the GST regime, finances of manufacturing and net producing states would come under strain, particularly in the initial years, till the cumulative benefits of better compliance can be realized in the form of higher revenues'.[12] These and other political bargains are inevitable in the transition to a new system; there was no doubt that the centre would have to guarantee that it would make up for any shortfalls in state revenue, and offer other incentives to those states which might need a little prodding to fall in line.

Even so, getting states on board and laying out policy guidelines—the tax structure, tariffs, the negative list of goods and services that would not be taxed under the GST, and so on—was not the most time-consuming and difficult part of getting the GST up and running. That distinction goes to the challenge of executing such a complex project across a country as diverse and as large as ours, exactly the same challenges we faced when it came to implementing the Aadhaar programme. Had the government waited until every policy decision was set in stone before moving on to the nitty-gritty of implementation, it would easily have taken years before the GST was ready for rollout.

Adopting the same philosophy of working in parallel that had helped fast-track the Aadhaar programme, Nandan declared to the task force, 'It will be foolish to wait for the laws to be passed before we begin thinking about implementation strategies. There's too much to do, and if we don't start right away, we could be looking at a five-year delay before the GST hits the ground.' Some of the immediate items on the to-do list were getting every dealer registered under a common nationwide registration scheme and designing the processes for operation and administration of the system. Software standards had to be specified, banks had to be integrated, commonly used tax software had to be upgraded, accountants and tax professionals had to be trained in the new regime, and the public at large had to be educated about the GST system, much like the VAT education melas that Tally had conducted earlier. To this end, Nandan proposed to the empowered committee that a Goods and Services Tax Network (GSTN) be created to do all this legwork in parallel with the legal discussions. The proposal was accepted unanimously, and in the Budget speech of 2012, the finance minister announced the formation of the Goods and Services Tax Network. Having an announcement by the finance minister with a date gave the team working on the creation of the GSTN the right amount of authority and importance to get the job done.

Getting down to brass tacks: The Goods and Services Tax Network

The GSTN is a new institution created specifically to act as the central technology platform that will implement the GST. The structure of the GSTN was articulated in great detail in the report of the Technology Advisory Group on Unique Projects (2011) chaired by Nandan, which drew on collective experiences from the execution of similar such projects in government.

Given that the power to impose and collect taxes is spread across the central and state governments, any attempt to build a centralized technology platform could be seen as a ploy to alter the balance of this power, creating instant opposition. One way to negate such potential dissent was as simple as choosing the appropriate name for the project. GSTN doesn't exactly make for a catchy acronym, especially for a project that aims to completely transform India's tax landscape. (In this case, Viral says it is he who is to blame—he wrote the report and ended up picking this name which eventually stuck.) However, there's more to the name than meets the eye; a neutral term like 'network' doesn't convey a threat, while terms like 'authority', 'India' or 'national' would imply a power shift towards the centre. 'Network' helped us convey our goal of the GSTN being only the plumbing that brought together all the stakeholders in the tax ecosystem, with no participation in any centre–state power struggles.

From day one, one of the biggest implementation-level concerns with the GST centred around the actual process of revenue collection. It was not clear to the states whether the centre would collect the revenue, and if the centre would share it with them promptly, or more likely late or never at all. Fund flows from the centre to the state have historically been delayed and intermittent, unable to provide much-needed support when state economies are struggling. It is this history that has led to today's complex taxation system, and has left our states deeply suspicious of sharing revenue collection powers with the centre. The loss of control over their own revenues was an unattractive proposition, and given that many states have tight finances, any delay in collections could adversely affect their functioning.

To address these concerns, the GSTN has been designed as a non-profit company jointly owned by the central and state governments, with professional management. By design, the government is not a majority stakeholder, allowing for the company to hire a professional team. The structure of the GSTN does not disturb the balance of power, and provides much-needed support to both the government and the people; revenue collection departments will find their work simplified through professional technology services, and people will avail of a customer-friendly set-up that transforms the adversarial relationship between the taxpayer and the authorities. It also grants a great deal of flexibility, allowing the organization to invest in choosing the right people for the task of building a complex, highly scalable technology platform. The creation of an institution such as the GSTN also solves another thorny issue—that of the taxation of interstate commerce. A common and neutral body that is jointly controlled by all stakeholders can settle interstate tax claims in a well-defined, timely and taxpayer-friendly manner.

While these decisions gave a sound footing to the fledgling institution, it wouldn't be government without its fair share of petty bureaucratic squabbles. Nandan strongly recommended that the chairman be drawn from the government, while the CEO be a professional from the private sector; he wryly recalls, 'One of the most heated discussions around implementing such a landmark reform was whether the heads of the organization ought to be from the Indian Administrative Service or the Indian Revenue Service.'

Technology-driven tax collection: Easy to comply, difficult to evade

Bharat Goenka narrates a story that makes clear the value of simple, easy-to-use technology platforms. 'Recently, the state of Karnataka launched a new annexure in their tax forms which required businessmen to provide invoice details. At the same time, they had a chat with us asking if this annexure could be uploaded from within Tally instead of the taxpayer having to do it manually. We agreed and launched the service in July 2014. In the first one month, the government expected no more than 3000 returns to be filed, thinking that people would need time to get used to the new system. To their absolute surprise, they got 30,000 entries.'

At the heart of the GSTN lies the technology platform that powers the implementation, operation and oversight of the GST. Coming up with a design for this platform was the mission of a group with another of those catchy government-issued names—the Empowered Group on IT Infrastructure on GST. Under Nandan's chairmanship, this group presented a report to the government in September 2010 which outlined the proposed system's design philosophy, key features and operational goals.[13] It would need a professional, highly experienced team to build a platform capable of processing hundreds of millions of transactions for millions of taxpayers.

One of the main aims of the GSTN is to make life simpler for both the taxpayer and the tax administrative authorities. As far as the taxpayer is concerned, the filing of tax returns and the payment of taxes should be a simple, uniform process, irrespective of location and the size of the taxpayer's business. Hence, the GSTN's technology platform should make it easy for the taxpayer to comply with the law without necessitating extra effort on their part. By implementing a uniform set of policies to be administered across the centre and individual states, the GSTN will also significantly reduce the costs and administrative machinery required to implement and enforce tax laws. The transparency conferred by the use of technology will help to plug leaks, eliminate tax fraud and enable easy auditing. Data from this system can be mined and analysed to improve tax collection. Ultimately, however, the GSTN architecture must recognize and respect the constitutional autonomy granted to the states, and should not blur the constitutionally drawn boundaries between state and centre.

Building the GSTN requires us to address the needs and concerns of all those who will be stakeholders in the system. Small taxpayers cannot be expected to immediately adopt an entirely electronic tax payment and processing system, and there should be extensive education and training to make the process as smooth as possible. Corporate taxpayers tend to operate across the country, and have to grapple with different tax regimes in different states. They usually have sophisticated inhouse software systems to handle the complexity of tax processing and payment. These systems must be made compatible with the GSTN by issuing a uniform set of standards to corporate tax software providers. These standards should also be implemented across all states, making it easier for state authorities to collect tax and implement policies. Tax collection, after all, is about actually collecting money, and hence the Reserve Bank of India and other banks should be able to freely work with the GSTN to get taxpayer information and process payments.

There are four key processes that the GSTN needs to implement, connecting the stakeholders we described above. The first is the registration of taxpayers. Since the PAN card is already a common ID used for paying income tax, it can be reused for the GST, without going to the pointless trouble and expense of creating a new, GST-specific ID format. Using the PAN number has the added benefit of reconciling direct and indirect taxes paid by every entity and creating a comprehensive record of every taxpayer's payment history, making it easier to audit payments and catch any instances of fraud.

The second important process in tax collection is the processing of challans. A challan is a payment instrument used to pay taxes to the government. Both the monetary payment and the associated challan are deposited at collecting banks, which then forward them to the respective tax administrations.

Finally, all taxpayers are required to file their returns so that the centre and states can both assess whether taxpayers have computed, collected and deposited their taxes correctly. These three processes—taxpayer registration, challan submission, and the filing of tax returns—can be completely automated and integrated with various tax-preparation software packages. Beyond these, there are various other processes such as processing of refunds, taxpayer audits and appeals, which also

TAXPAYER

Taxpayer pays tax using a challan

Taxpayer files returns to GSTN

HOW GST COLLECTION WORKS

VARIOUS GOVT-APPOINTED BANKS

GSTN PORTAL

MONEY COLLECTED

TAX REPORTED

FUNDS AND INFORMATION

CENTRE

Central and State governments receive their respective shares of funds and tax returns

STATE

need automation and integration with the revenue collection systems of the centre and states.

The fourth key process, and perhaps one of the most important and interesting ones, is the administration of the interstate GST settlement. Keeping in line with the destination-based taxation design of GST, in the case of interstate commerce, the GSTN computes how much tax each state owes each other, and settles the accounts of states. The entire process of GST collection and sharing of revenues is explained in the accompanying diagram.

While the GSTN can build this complex system that connects with all the stakeholders, defines all the interfaces, and provides a multitude of services to each stakeholder, the business rules—tax rates, negative lists, special categories and other such issues—can be decided independently by the government and simply uploaded into the system, and even changed as necessary. Policy-making and policy administration are thus cleanly separated, making the system far easier to operate.

Just as in the case of the state-level VAT, all buyers and sellers in the GSTN system will also be linked, raising the barriers to fraud and providing a significant revenue boost for the government. The GSTN can easily use data mining techniques to detect tax fraud—for example, the wrongful usage of tax rules or the creation of non-existent dealers to claim illegal benefits—and plug the leaks that exist in the current system.

Through the use of technology, the GSTN will tip the balance in favour of compliance rather than tax evasion, lowering the barriers for entry into the tax payment system while making it much harder to cheat on payments. This is the exact opposite of the approaches taken so far, where the emphasis is on vetting users before allowing them to pay tax, while fraud detection systems remain largely inefficient and vulnerable to cheating. Honest taxpayers will find the new system more transparent, cheaper, and far easier to use, whereas evaders will be caught more readily and suffer punishment. As we explain in the following diagram, the GSTN works as a self-policing system in which attempts at evasion are automatically flagged, making compliance easy

A SELF-POLICING TAX SYSTEM THAT INCENTIVIZES TRANSPARENCY

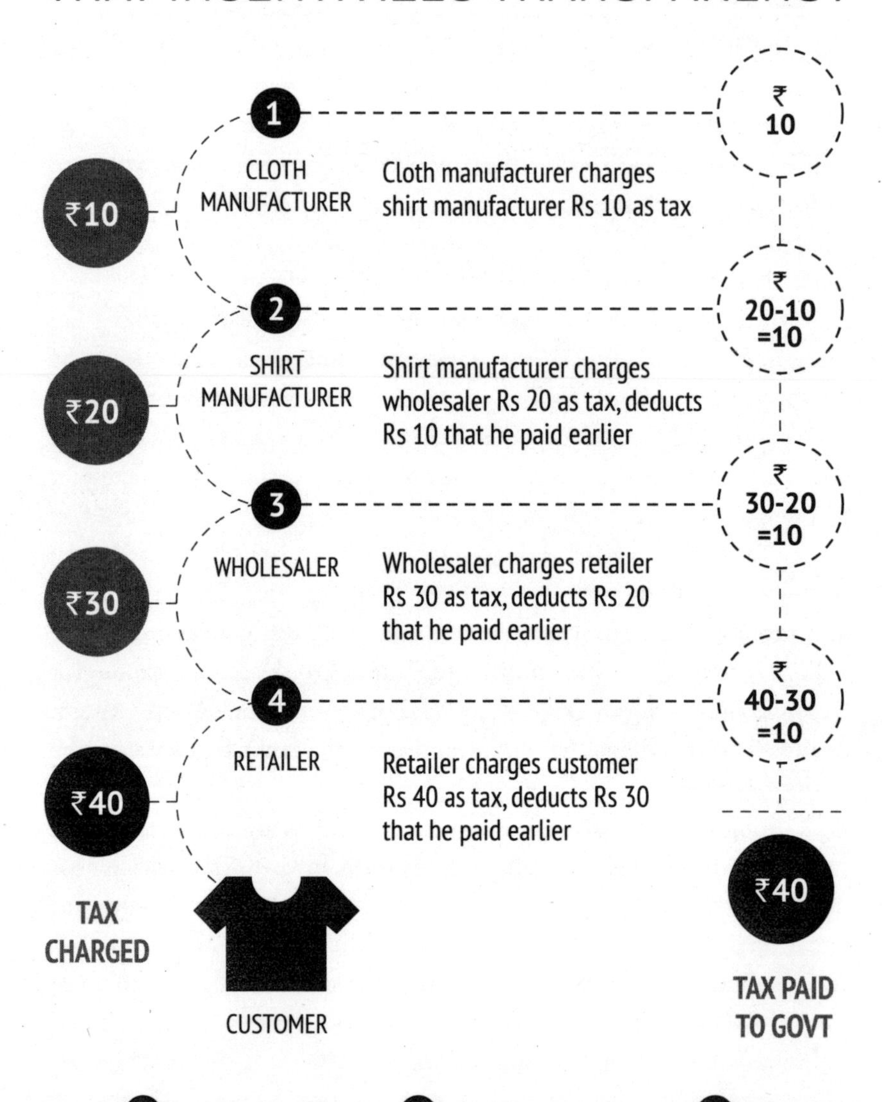

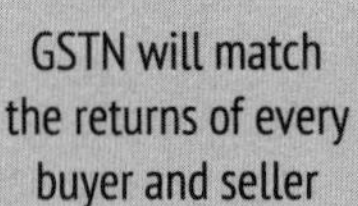

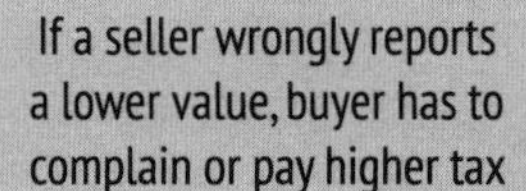

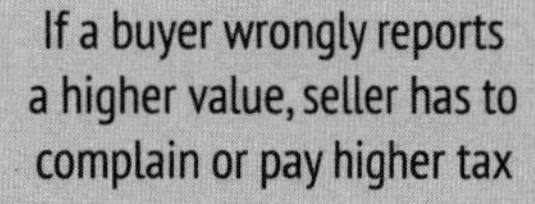

and our tax payment systems transparent. Implementing the GSTN can lead to an estimated 20 per cent increase in tax collections, which could work out to the tune of $4.6B (Rs 300 billion) in additional revenue for the government, enough to pay the annual bill of the nation's Rural Employment Guarantee Scheme.

The biggest tax reform since Independence

Creating a unified Goods and Services Tax is a critical step in building the kind of open market that will allow the Indian economy to reach its full potential, a step that Finance Minister Arun Jaitley has referred to as the 'biggest tax reform since 1947'.[14] We have an unprecedented chance to demolish several legacy taxation systems and streamline them into a single, uniform tax network. By simplifying taxes, we can spur a massive redistribution in India's manufacturing practices, where goods and services are produced where economical and consumed where convenient. All the wasted energy that goes into ensuring compliance with tax laws can be redirected where it is really required; technology can make compliance easy while simultaneously raising the barriers for evasion. Ultimately, the GST goes far beyond streamlining India's taxation network. It is in fact a project in nation-building, with the potential to radically transform the Indian market, the Indian economy, and 1.2 billion lives.

7

FRICTIONLESS HIGHWAYS FOR ECONOMIC GROWTH

ASK ANY INDIAN if he or she would like a better road network throughout the country, and the answer will be an instant 'yes', whether it's a farmer wanting a better way to transport his goods to the nearest market or a city dweller wanting to reach the office without negotiating spine-jarring potholes along the way. But ask the same person if they're willing to pay toll as a means to building this better road network, and the answer is suddenly a lot less emphatic.

In the absence of an open and honest debate about the need for toll collection, the entire issue has become heavily politicized. Tolls increase the cost of travelling by road, and using a tolled road for their daily commute pinches the pockets of the poor and the lower middle class; the resulting discontent is an easy target for those looking to gain political mileage and make the headlines. But reading between the lines, the fact that people have to pay toll is not the only source of anger; rather, it is the lack of transparency in the way toll systems are created and operated. Unfortunately, the focus of this rage often becomes the hapless tollbooth operator.

Terrified operators in Maharashtra fled as protesters descended upon a tollbooth outside Mumbai, breaking windows and smashing computers.[1] In Gurgaon, a driver was asked for his licence as proof that he lived in a nearby village, residents of which were exempt from paying toll; this request so enraged the driver that he shot and killed the tollbooth attendant.[2] Members of Parliament are also exempt from paying tolls; a member of Parliament from Gujarat who stopped at a toll plaza was asked to furnish his original parliamentary ID instead of a photocopy to avail of the exemption. He pulled out a rifle instead.[3]

Tolls are the price we must pay for using a modern, well-maintained road network that spans the nation. The need of the hour is to develop a toll collection system that is both effective and fair, creating revenue that can spur infrastructure development.

Bringing Kashmir and Kanyakumari closer to each other

Near Bhopal, the capital of Madhya Pradesh, lie two villages fifteen kilometres apart. One village is accessible by a proper road, part of the national road network; the other is not. Has this made any difference to the fortunes of their residents? In the first village, land is three times more expensive. People make 50 per cent more money, commute freely and tend small-market gardens whose produce they sell.[4]

This tale of two villages highlights the economic impact that better road connectivity can have, an impact that we have been historically slow to recognize. Roads are absolutely critical for the movement of people and goods across our country, accounting for about 85 per cent of all passenger traffic and 60 per cent of all freight traffic.[5] Despite this vital role, India's roads remained neglected for decades, with poor connectivity and even worse maintenance making road transport inefficient and painful. It was not until the 1990s that the government started paying serious attention to improving India's roads. Prime Minister Atal Bihari Vajpayee oversaw the establishment of the National Highway Development Project (NHDP), created to effect a massive overhaul of our national roadways; the initiatives implemented by the NHDP have been extended and expanded by all subsequent governments at the centre, a testament to the crucial role that a good road network can play in the nation's development.

The flagship of the NHDP was the Golden Quadrilateral Project, a national highway system designed to connect the four major metropolises of India—Delhi, Kolkata, Chennai and Mumbai. Once it was complete, the Golden Quadrilateral not only improved road transport across the nation, saving goods manufacturers time and money, but also ended up boosting industrial development and productivity in areas surrounding the highways. Upgrading India's roads could result in a savings of some $11B (Rs 750 billion) in fuel costs. For every million rupees the government spends on roads, 124 people rise above the poverty line; every rupee spent on building and maintaining rural roads yields more than Rs 5 in additional agricultural output.[6] Factories can now operate in formerly inaccessible and underdeveloped areas. One such example is Fabindia, famous for bringing Indian textiles and handicrafts to the global marketplace; through community-owned companies, they employ over 55,000 rural producers, and nearly 15,000 artisans are now shareholders.[7]

While the benefits of a well-built and well-maintained network of highways to boost connectivity across the country are extensive and well known, the unfortunate truth is that the government cannot single-handedly foot the bill for such large infrastructure projects. The only way we can build all the roads we need at speed and scale is through a partnership with the private sector, and with user charges. The revenues that the government collects in the form of toll can then be ploughed back into maintenance, building more roads, and helping scale up infrastructure development.

Quite apart from threats to life and limb, the current manual toll collection system suffers from other defects as well. Depending upon the agency collecting the toll, the receipt may or may not be computer-generated. Preparing receipts by hand is a slow and cumbersome process. Customers largely pay in cash, which forces them to carry enough money for the transaction, and requires toll attendants to count the money manually and hand over the appropriate amount of change. It is not uncommon to see an attendant from one booth jog over to another if his supply of small change has run low. The entire process is inefficient, slow, prone to leakage, and cannot scale up to handle increasing volumes of traffic. The National Highway Authority of India (NHAI) loses an annual $46M (Rs 3 billion)—15 per cent of the total toll revenue collected—in leakages.[8]

Consider the case of the Delhi–Gurgaon expressway, the busiest in India, used by over 200,000 vehicles a day and originally designed to transform a single-lane road hopelessly clogged with traffic into a world-class commute. Built under a public–private partnership, it was the site of South Asia's largest toll plaza, designed to allow the smooth flow of traffic. Instead, rush hour on the highway was memorably described as a 'gladiatorial contest', with commuters having to wait up to forty-five minutes to pay, thanks to a fatal combination of too much traffic, too few booths, and inefficient attendants. A shining idea clashed horribly with grim reality; the toll plaza is now closed, slated for demolition.[9]

Having to wait in long queues to pay toll worsens traffic congestion instead of lessening it, wastes fuel, reduces utilization, and works as a powerful disincentive for using tolled highways. Recognizing that a largely manual toll collection system was incompatible with the goal of building a modern, high-speed road network, Ravi Palekar, currently the CEO of Indian Highways Management Company Limited (IHMCL)—the organization responsible for electronic toll—was asked to design and develop an electronic toll collection system that could be implemented across the country. He recalls, 'One major challenge I faced was to decide which technology would work best. A different kind of challenge came from the people who were benefiting from the pilferage in the system. Any move towards modernization that brings in transparency and accountability always has its opponents, and I struggled quite a bit in the initial days.'

He eventually turned to Kamal Nath, then the minister for road transport and highways. 'I told him I was facing a big challenge in deciding which technology to use. He promptly called Nandan and asked him to contribute in this effort,' Palekar says. The government officially appointed Nandan in 2010 to chair a committee tasked with investigating existing technologies for electronic toll collection and recommending the most suitable option in the Indian context. This group included two outside experts in the field as well as officials from the relevant government bodies. Consultations were held with all the potential stakeholders in an electronic toll network: the NHAI, concessionaires, user groups and technology vendors. This is how government committees are—usually large, with all stakeholders and views represented. Successful committees bring about consensus across all stakeholders, balance all the incentives, and come up with recommendations that become the bedrock of implementation.

After evaluating multiple systems on such parameters as efficacy, cost, interoperability and vendor availability, the proposed design was based on radio frequency identification (RFID) technology, which allows for vehicle identification to become an automated process.[10] It is heartening to note that, as we discuss later, all the key recommendations made in terms of the technology for tolls and payment collection, as well as the organizational structure best suited for building a nationwide electronic toll platform have been adopted by the current administration and are being rolled out.

Zooming in: An Aadhaar for your car

Whether it's people or cars, we need a way to identify them reliably. People now have Aadhaar numbers to do the job; the equivalent for a vehicle is an electronic tag. How do such systems work in practice? As the accompanying diagram explains, vehicles equipped with an identity tag only need to slow down as they pass through the toll plaza. A scanner installed at the plaza reads the vehicle's details electronically, and automatically bills the toll amount to a linked prepaid account. There are several technologies that can be employed for this process—active and passive microwave, infrared, and active and passive radio-frequency identification (RFID). Microwave-based systems are common in Europe and Japan, whereas RFID-based systems are popular in the US.

Based on the technical merits and costs of different technologies, the committee chose passive RFID as the best solution for India. In this system, the vehicle owner needs to purchase a radio tag and stick it prominently on the windscreen. The tag has an antenna and

EXISTING TOLL COLLECTION CHALLENGES:

Long wait times

Cash transactions

Opportunities for diversion

ELECTRONIC TOLL SYSTEM

LANE 2 | LANE 3 | LANE 4

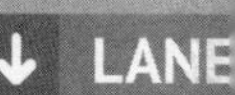

Vehicle goes through gate without stopping

Electronic tag is read

Account of tag holder is debited

Account of toll-collecting agency is credited

BENEFITS:

HIGHER COLLECTION

LESSER FRAUD

SAVES TIME

SAVES FUEL

NO CHANGE HASSLE

6
EASIER MOVEMENT OF GOODS

a circuit printed on the inside; they are activated by the overhead reader, or transceiver, installed in the toll plaza. Such systems are now commonplace and the cost is minimal. In the passive RFID set-up, the main source of energy is the overhead reader, not the sticker. In practical terms, this means that the vehicle owner does not have to bother with changing batteries or ensuring that the tag on his car is charged. Once the sticker is in place on the vehicle, it requires no maintenance on his part. The reader in the toll plaza can easily read the tag as the vehicle passes underneath, reducing the transaction time to a fraction of what it is currently.

For any system that is likely to be deployed nationwide, the technology component must be robust, tried and tested in the real world, and affordable. Ideally, it should be available from multiple vendors so that the government is not forced into a lock-in with a single vendor. For the RFID-enabled toll collection system, there need to be multiple vendors of tags and transceivers, all developed to a common specification laid down by the government. Once the specifications are met, any tag should be capable of being read by any transceiver and vice versa, irrespective of the vendor. The passive RFID technology satisfies all these criteria, and hence was recommended by the committee as the technology platform on which to build an electronic toll collection system.

Zooming out: Electronic toll collection

Once you have set in place the method to identify vehicles, you need to design a standardized process for collecting toll. What are the steps involved? First, the tag on a vehicle needs to be linked to a prepaid account, much in the way a prepaid SIM card works. Every time the vehicle passes through a toll plaza, funds will be debited from the account, and the tag must then be topped up in the same way that we top up our prepaid mobile connections.

Ravi Palekar tells us, 'When the user registers with the bank, they have to pay a one-time cost for the tag. Right now, that cost is around $4 (Rs 250) plus tax, but that number will decrease significantly as the user base grows. A minimum balance has to always be maintained on the card. This threshold has been set at the amount of toll applicable to that class of vehicles for 120 kilometres, roughly the cost of passing through two toll plazas. The user will keep getting SMS alerts to top up their account when the balance is low, to avoid being blacklisted.'

As toll plazas become more efficient, transport operators can track fleet movement, and toll road users can drive non-stop without bothering about toll rates or carrying exact change. The same tag can be read at toll plazas on highways all across the country so that state boundaries no longer act as deterrents to the smooth flow of traffic. Coupled with the tax reforms we discussed in the previous chapter, we envision a situation in which the same state barriers that today impede the flow of goods and passengers will become invisible for all practical purposes. We will have a single uninterrupted pan-India road network, increasing connectivity and bringing even the most far-flung regions of the country onto the national grid. Trucks will no longer need to prominently display the names of the states in which they have a permit to operate, which can be encoded on the tags instead. There'll be plenty of room for the colourful artwork and the catchy slogans—like the classic 'Horn OK Please' and the admonitory '*Buri nazar wale, tera moonh kala*' (O you with the evil eye, may your face turn black)—that make our trucking fleet such a memorable part of India's roads.

The electronic toll payment system, in its broad outlines, is quite similar to the telecom prepaid model. The three main components of such a toll network are the distribution of tags, the registering of users and the actual transactions themselves. Given that the tags are low-cost and easy to transport, it is possible to imagine a large retail ecosystem where they are available for sale. Petrol pumps and service stations are frequently visited by vehicle owners, and it makes the most sense to stock tags in these locations, alongside the fizzy drinks, packaged snacks and cans of motor oil; ultimately, tags would be available for sale at most retail stores.

The next step is the creation of recharge points, where funds can be loaded onto the tag. Just as mobile phone users can top up their prepaid accounts anywhere from a swanky mobile showroom to a corner stall, so also tag owners should have a multitude of options to top up their accounts. Eventually, tag recharge should also be possible online, using the kind of universal electronic payment system we have discussed in an earlier chapter.

Today, top-ups can already be done at toll plazas. Ravi Palekar tells us that truck fleet owners have been quick to embrace the electronic toll system for the convenience it offers. They no longer have to issue cash to their drivers and can easily recharge every vehicle's balance from the comfort of their office, while drivers save time and fuel by avoiding the wait to pay tolls in cash.

How would such a toll collection platform be managed? The committee that studied this problem recommended the creation of a Central Electronic Toll Collection System (CES) that would link all the players in the system—the seller issuing the radio tag, the user loading funds to their prepaid tag, the toll plaza operator managing their concessionaire account—while also providing reports to the government. The CES would manage the accounts of all the stakeholders and at the same time handle a number of other processes, such as the tag issue and recharge system and processing of transactions. Toll plazas are already being upgraded to use tag readers and are being connected to the CES network.

For electronic toll collection to make sense, it must be fast-paced—a vehicle cannot spend minutes idling in a toll plaza while the toll transaction is processed online. To make the process as quick as possible, every toll plaza should be linked to the central system via the internet. A vehicle's tag should be scanned, the balance on the account checked, and the appropriate amount debited in the few seconds it takes for a vehicle to move through the plaza; all these transactions must be recorded in the CES. Palekar explains, 'In reality, connectivity is a challenge since toll plazas are often in remote areas. The toll plazas actually synchronize with the central platform every half an hour, uploading a list of all vehicles that passed and the toll to be charged, and downloading a blacklist of offenders. As connectivity improves, these transactions can happen instantaneously and online.'

The success of this entire system depends upon reliability and convenience. The concessionaire who operates the toll plaza is responsible for maintaining the equipment that scans the vehicles as they pass. The reliability of payment processing, from easy recharge to toll payments, is equally important. Banks will handle the financial aspect of toll collection; in order to encourage competition and build robust payment systems, it is essential that multiple banks be allowed to participate. As of now, ICICI and Axis Bank are the two participants, and more can be added as the toll system grows.

IMHCL's tag design accounts for the fact that more banks are expected to join the system. Each tag has a number, the first two digits of which are reserved for identifying the bank that issued the tag. Thus, if a vehicle owner got a tag from Axis Bank, when he passes through a toll plaza, the reader will note the first two digits, identify the issuer and then send the transaction to Axis Bank to be processed. The bank will check the balance on that tag and deduct the amount from the tag holder's account.

There are, of course, possibilities for potential fraud. Even in a situation where all vehicles need to carry an appropriate tag, people might still try to game the system. For instance, a car or truck owner might tag their vehicle with a motorcycle sticker so that they have to pay a lower fee. People may also try to just zoom past the toll plaza without paying. Systems are being put in place to ensure that these types of fraud are eliminated. Toll plazas are equipped with cameras that can read licence plates and record offenders as they speed by; this is also a backup in case the tag reader malfunctions. The entire electronic toll collection process has been designed to make it easy and quick to pay, while making evasion difficult and punishable by law. Such a strong monitoring system with the appropriate legal backing serves as a deterrent for potential evaders.

Where rubber meets road: Electronic toll today

The central government has decided to officially launch the Electronic Toll Collection (ETC) network nationwide. A hundred and sixty out of 340 toll plazas have already been linked to the ETC system. Pilot projects have been carried out on the highway between Mumbai and Ahmedabad, and the ETC system has been installed in fifty-five toll plazas along the Delhi–Mumbai route, making it a seamless ride without any stops along the way. The Mumbai–Chennai highway is next on the anvil for seamless connectivity. In addition, the government has declared that all future highway projects must have a provision for ETC lanes.[11] Based on a study conducted by the Indian Institute of Management, Calcutta, and the Transport Corporation of India,[12] the ETC system could save the economy an estimated $13B (Rs 870 billion) annually in productivity that is lost at toll plazas.

Coming back to earth, the two participating banks, ICICI and Axis Bank, are part of the official electronic toll programme slated to operate along the Delhi–Mumbai highway.[13] Chanda Kochhar, managing director and CEO of ICICI Bank, was quoted as saying, 'ETC substantially enhances convenience for users, and we believe it will play a very important role in contributing to the growth of cashless payments in India.'

IMHCL is the organization tasked with establishing and maintaining a national electronic toll collection network, and its structure is similar to that of the GSTN which we discussed in the previous chapter. It has been designed as a collaboration between the NHAI, the private concessionaires who are tasked with toll booth operation, and financial institutions who handle the payments process. This design, in Ravi Palekar's words, 'bring[s] in the best of both worlds—good practices from the private sector and regulatory control and oversight from the government'. The NHAI holds 25 per cent of the company, financial institutions hold 25 per cent, and the remainder is held by concessionaires. In this way, the government retains strategic control without getting bogged down in the minutiae of operation and maintenance. These are the same thoughts that guided the establishment of the GSTN, and it is heartening to see the cross-pollination of these ideas in other domains.

Thinking big: An ecosystem beyond just toll

Once every vehicle carries a radio tag linked to a payment account, we now have a national network that can be used for many things other than just toll payments. As Palekar puts it, 'The tag becomes a unique identifier for the vehicle, and can be used for many applications beyond toll alone. It works as an automated vehicle identification system, and the ETC is just one application of that.' A prepaid toll system that works across banks and spans the entire nation can be used for other types of transportation-related payments. This could include paying for local tolls in states and cities, parking charges, fines for traffic violations, road and vehicle tax, and any other transaction requiring payment from the user.

Another extension of the ETC system can be towards reducing traffic snarls by levying a congestion tax. In such cities as London and Singapore, users are charged a fee for operating a vehicle in some areas during peak hour, helping to reduce the traffic burden.[14] The best we have managed in India so far is to ban commercial vehicles from entering urban areas during certain hours. In Mumbai, for example, trucks are not allowed to ply inside the city during the day, and can only enter at nightfall; however, this system doesn't extend to all major cities in the country. Being stuck in a typical peak-hour Bengaluru traffic jam, surrounded by vehicles ranging from sputtering scooters to ramshackle trucks groaning under the weight of rocks or sand destined for construction sites, is sufficient to convince us of the value of a congestion tax in making everyday travel a less nightmarish experience. Transceivers mounted at the appropriate locations—the city limits or the borders of the business district—can be used to scan the radio tags of vehicles entering the area, and users would then be charged a congestion fee electronically. Alternatively, if the decongestion approach involves banning the entry of vehicles except during a specified time period, the RFID system could be used to flag offenders and levy fines.

Today, vehicles have to stop at state borders to hand over their manifests and undergo inspection, which wastes time and lowers transport efficiency. Truck delays at checkpoints cost the economy an estimated $138M (Rs 9 billion to 23 billion) in lost operating hours. A perverse economy has sprung up in which truck operators pay bribes to officials to avoid paying taxes on their goods, or fines for overloading their vehicles. A World Bank study estimates that these 'facilitation payments' can run anywhere from $138M (Rs 9 billion) to $1.1B (72 billion). The net result is that the state loses money due to it in the form of tax. Officials overlook violations of traffic safety laws, and this contributes to even larger economic and social losses from accidents that could otherwise have been prevented. The annual economic loss from road accidents is estimated to be in excess of $8.5B (Rs 550 billion), and the majority of these accidents involve trucks.[15]

With the ability to identify and track vehicles through an online system, the electronic toll platform could be used to create digital manifests for goods transport. A user, say a trucking company, would voluntarily sign up to register the tags of all their vehicles on an electronic manifest system. Every time a vehicle is dispatched, the company would create a manifest of the goods being carried, especially relevant in interstate transport where there are restrictions on the goods that can be moved across state borders.

With an electronic manifest system, the details of a vehicle's cargo would be available automatically as the vehicle passes through a transceiver-equipped inspection station, speeding up the process considerably. A similar system could also be used for vehicles entering city limits, as well as lending itself to the computation of octroi or other state- or city-level taxes on transport. An electronic manifest system would help decongest state borders and remove one more obstacle in the path to making road transport in the country efficient, effective and hassle-free. Even more importantly, automating the tax computation process would eliminate incentives for corruption, resulting in a significant increase in revenues and a gradual improvement in road safety standards as well.

Ravi Palekar explains to us, 'If you integrate the RFID and taxation systems together, every time you read a tag at a check post, you can get the vehicle's entire history—its origin, destination, what goods

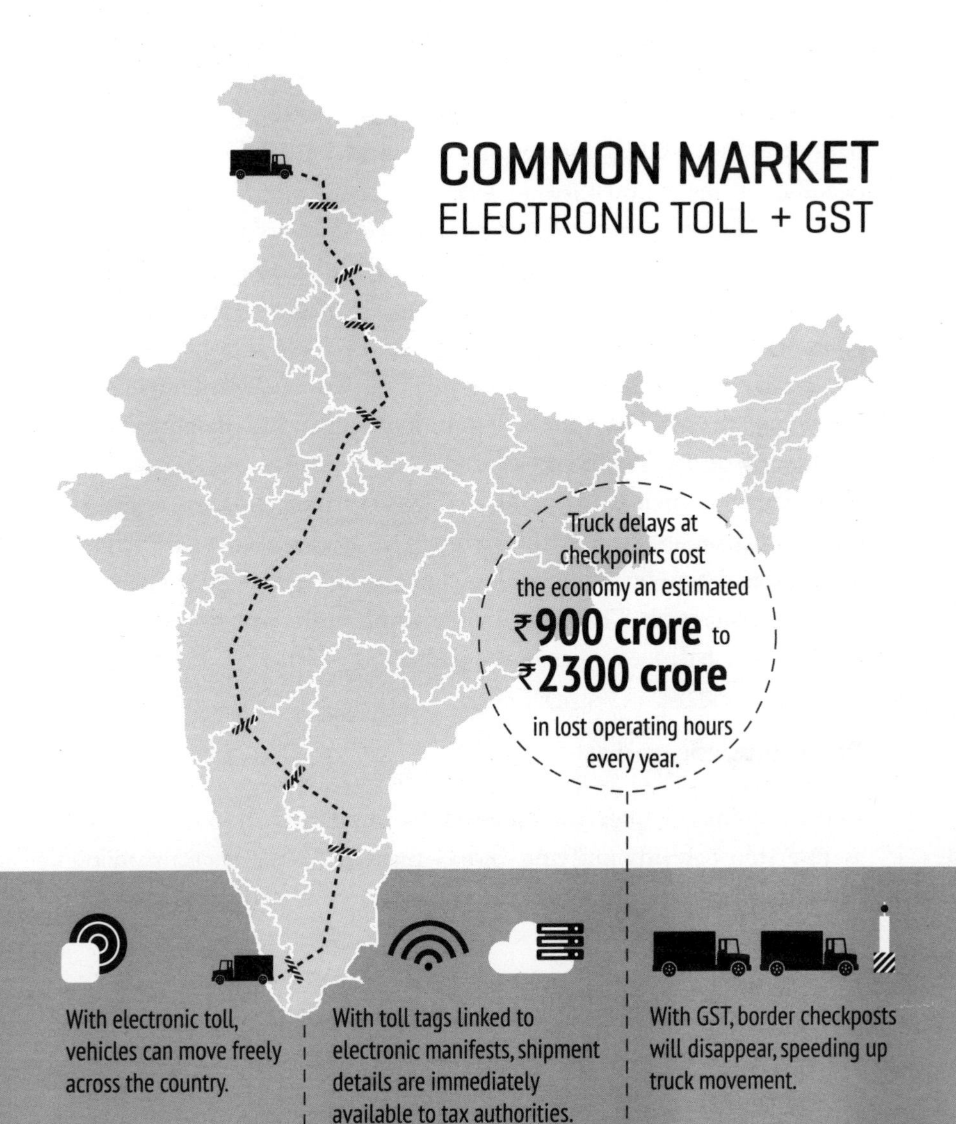

With electronic toll, vehicles can move freely across the country.

With toll tags linked to electronic manifests, shipment details are immediately available to tax authorities.

With GST, border checkposts will disappear, speeding up truck movement.

IMPACT:

By creating a single common market, electronic toll and GST will provide a huge boost to India's manufacturing sector and the 'Make in India' initiative. Just-in-Time manufacturing and lower fuel and transportation costs will increase efficiency.

it's carrying, whether the operators have paid tax or not—in a format that's easily available to the state.' A separate steering committee is working on integrating the radio tag into the state border check-post system, and he says, 'A team will shortly be visiting Ahmedabad to consult with the Gujarat government about taking the RFID-border check-post linkage forward.'

Once the government develops 'electronic toll as a platform', not only will various government applications that we discussed above be possible, but also, start-ups will come up with all sorts of innovative ideas once they have a method to identify a vehicle and the ability to debit and credit its wallet in the cloud. This will necessarily bring up the issue of privacy, but it is worth pointing out that it is straightforward in today's world to electronically read a car's licence plate as it passes by, and companies such as Ola and Uber already have significant amounts of information about our movements. This is yet another case where the potential benefits of technology have to be balanced with privacy and data protection. The government has an important role to play in defining the balanced rules of the game.

One nation, one market

In the previous chapter, we discussed the implementation of the GST as one step towards unifying India's market. As the accompanying diagram explains, removing transportation roadblocks through the implementation of an electronic toll collection system would help to build a true common market, where goods could be freely transported all across the country. Vehicles would not need to halt at any checkpoints along the way, since the list of goods they are carrying as well as the tax that applies to them can be calculated and billed automatically.

At a time when the government aims to make India a global manufacturing hub through its 'Make in India' programme, the combination of electronic taxation and toll systems can be a powerful launching pad for such initiatives. Whether it's improving traffic in cities or creating an efficient road transport network across the country, simply building more flyovers or more highways won't solve the problem. Creating intelligent systems that use technology to flatten out the biggest bumps in road transport, and then applying these systems in innovative ways to solve other transport-related issues is the only way in which India's investments in infrastructure development can bear fruit. We've already seen the positive economic impact of bringing villages into a national road network; imagine the possibilities of a system that could connect every corner of the country in one fell swoop.

8

STREAMLINING GOVERNMENT SPENDING

> Just as you cannot tell whether a fish swimming under water is drinking water or not, so also you cannot tell whether government servants employed in government work are taking money or not.
>
> —Kautilya, *Arthashastra*

SOCIAL WELFARE RECIPIENTS make for unlikely owners of luxury cars. Yet, an investigation by the Brazilian government's Public Spending Observatory initially indicated that many beneficiaries of the Bolsa Familia scheme were also listed as owners of high-end vehicles in the federal car registry. With a little bit of digging, this unusual statistic was found to be the result of a tax dodge; the real, and considerably wealthier, owners of these vehicles were registering them under the names of Bolsa Familia recipients to avoid paying taxes.[1]

We've already read about Sanjay Sahni, the Muzaffarpur electrician who uncovered instances of fraud in the government's rural employment guarantee scheme. All he did was to find an official list of beneficiaries from his village on the internet; some of those on the list hadn't worked for the scheme at all, and others had not received a full payment. Instead, the money was going to middlemen and village officials. Sahni was able to mount a successful campaign so that his fellow villagers could get their rightful dues.

Let's say you are a taxpayer who happens to be curious about how your tax money is being spent. If you live in the United Kingdom, you can go online, enter your annual income and find out not only how much tax you pay yearly, but how that tax money is being used every day—whether it's in running the government, funding medical research or in welfare programmes for the elderly and disabled.[2]

As the examples above illustrate, citizens have the right to know where their governments are spending their money, while governments themselves can use a better understanding of their finances to weed out fraud and get more bang for their buck. The idea of transparency in government expenditure is gaining traction around the world, and a coalition of nations has now come together under the umbrella of the Open Government Partnership (OGP), a global initiative launched in 2011.[3] The OGP engages with governments to promote transparency and citizen empowerment while eliminating corruption, and advocates the use of technology to improve governance.

Many member nations have launched federal transparency portals to report on government finances and expenditure, in keeping with the principles of the OGP. Brazil's transparency portal, for example, has provided private citizens with an unprecedented level of access to keep tabs on the flow of government money in a country as riven with corruption as our own. The Brazilian government itself routinely analyses data from the portal and other internal databases to monitor spending. Journalists use the portal to uncover individual instances of wrongdoing, as in the case of a minister who inappropriately made the government foot the bill for his personal expenses. In an example of how such open platforms can be further leveraged, Brazilian programmers have created a website that takes information from the portal and presents it in a simplified format, making it easy for users to understand the details of public spending in such areas as education and transportation. Other countries, such as the US and the UK, have also created large online repositories for the government's financial data.

Where does the government's money go?

The value of a transparency portal in streamlining India's government expenditure is unquestionable. As an illustration, consider the fact that well over a thousand social security schemes are operational in the country today, from the centre down to the state level. Yet there is no single, uniform and user-friendly system that provides a clear outline of exactly how much money was allocated to each scheme and on what it was spent. The information you can find tends to be heavily outlay-driven, giving only an overall picture of the monies allotted to a specific programme; there is no detailed breakdown of expenses, there is no link to a set of specific outcomes; it is quite likely that any available data is a few years out of date and, most crucially, no data is available below the district level, in effect cutting off the flow of information just as it reaches the individual citizen. Much of this material also tends to be scattered across multiple government reports, making it even harder to get an integrated picture of government spending across the country.

Without up-to-date information on fund utilization and outcomes, it becomes difficult for the administration to evaluate the effectiveness of its public-spending programmes and to decide upon a future course of action. In the absence of such data analysis, we end up judging programmes on outlays rather than on outcomes. Lacking a mechanism to track the disbursement of funds to implementing agencies, we are left with no way of measuring their performance. Agencies themselves suffer when funds do not reach them in a timely fashion, either due to lack of information about when money enters the system or because of diversion. Funds also tend to float unused in the system, increasing inefficiency and waste. Since the flow of information is top-down, delays are inevitable, further muddying the waters. In the absence of a central system, agencies tend to implement their own mechanisms for handling fund flow and monitoring, resulting in a slew of individual systems that are incapable of talking to one another. As Yamini Aiyar and T.R. Raghunandan from the Accountability Initiative put it, 'One of the necessary conditions for ensuring outcomes against public spending is that expenditure be aligned with needs and priorities on the ground. But the late arrival of money coupled with the Indian bureaucracy's passion for paperwork (in this instance, filling up utilization certificates), makes for a deadly cocktail of wasted expenditure.'[4]

In India, we already have several institutions working with publicly available data, performing research and compiling statistics in a user-friendly format for public consumption. These include the data-driven journalism portal IndiaSpend, private research initiatives like the Accountability Initiative and the Centre for the Monitoring of Indian Economy and the publicly funded National Institute of Public Finance and Policy. Much like the Brazilian model, the government should focus on releasing all its raw data related to expenditure, and journalists, researchers and public interest groups like the ones we list above can analyse this material to uncover relevant trends and provide public updates.

Opening up the books: The Expenditure Information Network

In 2011, the ministry of finance set up the Technology Advisory Group for Unique Projects (TAGUP) to study how technology could be incorporated in five large-scale financial sector projects.[5] Given the urgent need for a streamlined expenditure management system in the country and the growing success that federal transparency portals—the US government's data.gov initiative is a notable example—have enjoyed in other countries, one of the five projects was an Expenditure Information Network (EIN). Similar in spirit to a federal transparency portal, the EIN will allow a citizen to trace how their tax money is being spent from the centre all the way down to the local level. The Right to Information Act (2005) already mandates that every public authority must voluntarily digitize all appropriate records and enter them into a nationwide system for easy access; they must also voluntarily release certain specific types of information related to their structure, function and performance, making it freely available to citizens without the need for a formal request to be filed.[6] The EIN is therefore a concrete implementation of the spirit of information sharing espoused by the RTI Act.

The EIN has been envisioned as a single integrated network that will function as a transparent accounting and auditing system for government expenditure at all levels, from the centre to the district and beyond. The EIN will allow for the monitoring of various government schemes both in terms of funds utilized and outcomes generated, information that can also be used to evaluate the implementing agencies. The effectiveness or lack thereof of a particular scheme can then be used to drive informed policy decisions. Funds can be disbursed quickly and efficiently, and their usage can be tracked in near-real time and monitored till the last mile. This network also lends itself to the creation of a federal transparency portal which will empower public scrutiny of government spending.

Of particular note is that the EIN was also envisioned as a thin solution, a technology platform that brings together all stakeholders without disrupting existing systems and controls. Agencies will retain control over activities like approvals and sanctions, and will be granted the freedom to integrate the EIN platform into their existing programmes at their own pace. Those expected to become part of this network include administrative policymakers, government programme divisions within specific ministries, central monitoring agencies such as the Controller General of Accounts (CGA), state-, district- and village-level administrators, the agencies who actually implement various schemes on the ground, beneficiaries, experts, and of course, the general public.

Much like the set-up of the GSTN, the best organizational structure for implementing the EIN would also take the form of a National Information Utility (NIU). The NIU structure will also be better suited to partner with government bodies across various administrative levels. This is a critical requirement given that other expenditure monitoring agencies, such as the CGA, operate only at the central level. Since much of the expenditure happens at the grassroots where the writ of the CGA does not extend, an EIN must necessarily be housed in a neutral body which can conduct business with central-, state-, district- and even village-level administrations.

The first step to building an EIN ecosystem is to register all players on a common platform. The details of every programme will be captured on this platform, starting with the programme definition. Fund allocations, fund flow and utilization will be tracked through the system, and feedback can be solicited as needed. From the point government money enters the system till the point it actually reaches the citizen on the ground, every step in the process is transparent and traceable.

In the existing set-up, information flows only one way through the system, from the time it enters at the top. The people on the ground have little to no way of providing feedback to the system, making real-time monitoring virtually impossible. We envision the EIN as an entity set up along the lines of a hub, with every participating entity being a spoke. Unlike a linear model, the hub-and-spoke system allows for the free flow of information among entities, whether it's the central government or a gram panchayat. Delays and blocks can be flagged as soon as they arise; since every player in the system is aware of the problem, the impetus towards a speedy resolution is all the greater. The hub-and-spoke model also allows for asynchronous rollout, as we discussed in the case of Aadhaar, where different government bodies come on board at their own pace.

How transparency helps: Tracking India's education expenditure

Let's explore some of the problems with government expenditure by turning the spotlight on education. How does a government-run primary school in India spend its money? Which is a higher priority—buying chalk, repairing toilets and drinking water facilities for students, or whitewashing the school walls? The answers to these and other questions can be found in the 2012 Planning, Allocations and Expenditures, Institutions: Studies in Accountability (PAISA) report, 'Do Schools Get Their Money?', produced under the aegis of the Accountability Initiative, a research organization;[7] the data from this report, in our opinion, make a compelling argument for the role an EIN could play in ensuring that government spending actually meets its intended target. PAISA, which is India's only citizen-led effort at the national level to track government spending on education, surveyed nearly 15,000 government-run elementary schools across the nation, and asked some simple questions, such as: 'Do you get the money allocated to you from the government, and if you do, when and how much of it do you get?'

Government schools get money under the Sarva Shiksha Abhiyan (SSA), meant to promote universal elementary education, and this is further divided into subcategories for specific use—to maintain school infrastructure, for operational and administrative costs, and for obtaining instructional aids for the classroom. The survey found that of the three, the second category was the only one where every government school could be sure of receiving the allocated sum in its bank account—and that's only 2 per cent of the total money the SSA scheme disburses. Schools had little control when it came to deciding how to spend their funds, since these decisions are controlled at the central level by the state government. So, as it happened in Himachal Pradesh in 2011, if the state government orders all the schools in one particular district to build boundary walls, they must comply, regardless of whether this is a fruitful use of their limited funds or not. Similarly, if the chief minister decides to pay a visit, schools are forced to whitewash their walls, or buy cupboards and other furniture that may not be their topmost priority.

Halfway through the financial year, the survey found that only about 50 per cent of schools had received their grant money; the official explanation for this delay included limited access to core banking facilities, banks not transferring funds on time, wrong bank account numbers, and confusion around the names of the grants. Here is an immediate problem that an EIN could fix. An electronic system for managing fund flows would ensure that all schools received their money on time.

Far more schools—67 per cent, to be exact—spent their money on whitewashing their walls rather than repairing toilets (36 per cent) or

^ A village kirana shop, April 2010. Millions of such stores dot India's landscape and, equipped with microATMs, can bring banking facilities to every Indian's doorstep.

Photograph © Naman Pugalia

^ Aadhaar logos being tested in a village in Azamgarh district, Uttar Pradesh, in April 2010. A competition was held to design the final Aadhaar logo and various options were tested in the field before selecting the best design.

Photograph © Naman Pugalia

^ Nandan at UIDAI's Bangalore tech centre, August 2010. The pictures on the wall behind him were taken by team members travelling across the country, serving as a constant reminder of both India's diversity and the size of the challenge we faced.

Photograph © Naman Pugalia

^ A wall in Tembhli village, 2010, on which data from a government scheme is being painted. This is the local transparency portal, where residents can check the data for accuracy.

Photograph © Naman Pugalia

^ Viral in Hazaribagh district of Jharkhand, 2011. He was configuring the microATM device that would be used in initial field trials, and attracted a crowd of fascinated children.

Photograph © Naman Pugalia

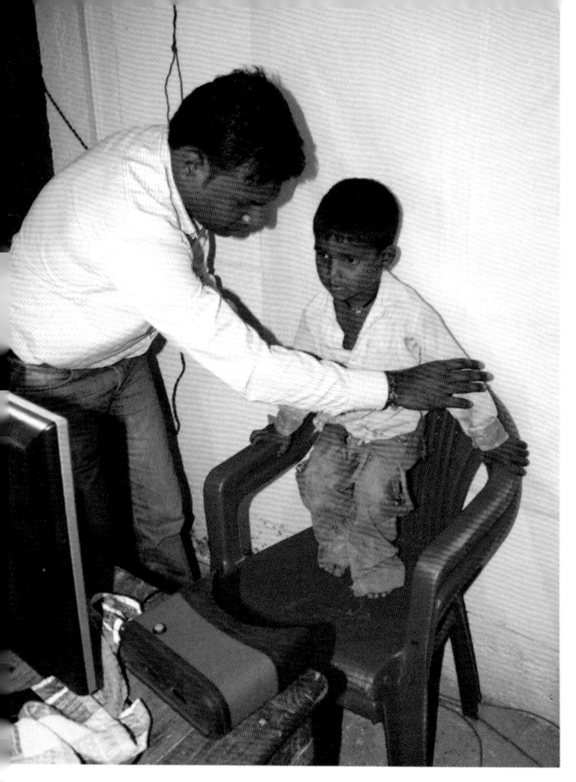

< A young resident of Tembhli village being readied for Aadhaar enrolment on 20 September 2010, just prior to the official launch of the scheme. Even though the enrolment technology had been standardized, such makeshift arrangements had to be made on the ground to accommodate everyone.

Photograph © Naman Pugalia

^ The first line of residents waiting to get enrolled in Tembhli village, 20 September 2010. As enrolments scaled up across the country, such lines became commonplace.

Photograph © Naman Pugalia

^ Aadhaar letters being delivered by a postman in a Jharkhand village, February 2011. The book in the postman's hand is a special register created by India Post to track the delivery of Aadhaar letters.

Photograph © Naman Pugalia

A typical rural employment worksite, May 2010. These women are waiting for the muster roll call to mark their attendance under the MGNREGA rural employment guarantee scheme.

Photograph © Naman Pugalia

Posters on a wall in Khurja, Uttar Pradesh, April 2010. These advertisements reveal India's aspirations—classes teaching computer programming and how to speak American English hold the promise of better jobs and more money for India's youth.

Photograph © Naman Pugalia

^ Posters advertising entrance examinations for various government services in Uttar Pradesh, April 2010. A government job is usually seen as a guarantee of a secure income and a comfortable life.

Photograph © Naman Pugalia

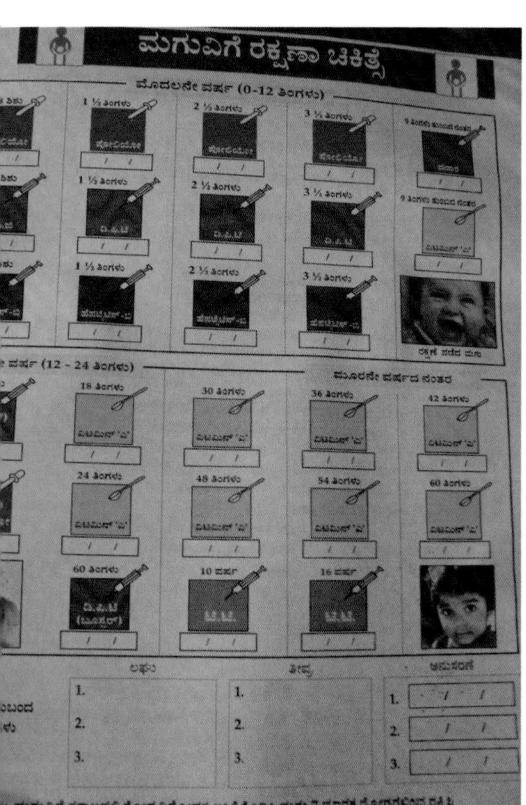

^ A public health centre on the outskirts of Bengaluru, May 2010. The computer at this health centre is draped in protective covers and kept under lock and key in a separate room, and is of little use.

Photograph © Naman Pugalia

< At the same Bengaluru primary health centre, while the computer languishes unused, paper records like this one are used to track the vaccinations administered to a child.

Photograph © Naman Pugalia

on drinking water facilities (44 per cent). Does this mean that having pristine walls is more important to schools than providing usable toilets and clean water to their students? Not really; it's just that a whitewashed wall is an easy, tangible outcome to point to when funds have to be spent quickly. What this also indicates is that expenditures are driven more by the diktats of the government and don't always connect with the need on the ground. Here is another area where an EIN could bring in much-needed change. By simply being able to track in real time where schools are spending their money, the administration can immediately identify and hopefully eliminate mismatches like those above so that schools are given funds to address what they genuinely need. By bringing government, schools, banks and the public onto the same platform, an EIN can bring about a shift in the current mindset that prioritizes outlays over outcomes. As the report puts it, 'Unless expenditures are targeted towards efforts at translating schooling into learning, the increasing public expenditure on elementary education will be wasted.'

However, an EIN cannot fix some of the fundamental issues around the education sector, issues that crop up in many other areas as well. For example, it cannot resolve the problems around the centralization of education, where schools are run from New Delhi or from the state capital. It cannot solve the problem of absentee teachers, or the poor quality of midday meals. We can only hope that complete transparency will pave the way for decentralization, where communities participate actively in their schools. Today, if alternatives are available and the family can afford it, they will immediately send their children to a private school.

We believe that centralization arises from opacity. Within different levels of government, officers can allocate funds but do not know how these are spent. This lack of information prompts risk aversion to kick in so that more rules spring up to safeguard the usage of government money. With greater transparency at all levels, we believe that money management can be decentralized and the common man on the ground can have his voice heard by those in the corridors of power.

The three laws of bureaucratic evasion

In the years since the proposal for an EIN was first made, there has been very little progress towards implementation. In fact, Nandan had made a similar suggestion as far back as 2008, and related proposals have been mooted over the last seven years without much success. Our own efforts in this direction were stymied by what we have come to call 'The Three Laws of Bureaucratic Evasion'. These laws have arisen from the bureaucracy's emphasis on stability, and its discomfort with new ideas, as well as the fear that it will reduce power and control among the established players. These apply not just to the EIN but to Aadhaar and to any scheme which proposes to 'initiate a new order of things'.

The first law is: 'We don't need this.' It's too expensive, it does not fit the Indian 'culture and ethos', it's irrelevant to our problems or of dubious benefit to the people, it will not work, it will upset everyone, this is not the way we have done things—innumerable justifications can be found as to why the people of India do not need whatever it is that you are proposing to do. The second law: 'Why are you proposing this? It belongs to some other department.' In other words, you are treading on someone else's toes and interfering in a project that some other agency is in charge of. And the third law: 'It's already being done!' At any given time, there are dozens of projects floating around the corridors of government in various stages of completion, and any one of these can be pulled out to demonstrate that what you plan to do is already being implemented by someone else. Of course, many of these projects never actually reach closure, serving only as handy deterrents when needed.

Despite these setbacks, we remain convinced that building an Expenditure Information Network is an essential step towards reforming and modernizing India's economy.

A road map for transparent government expenditure

How do we see the future of government expenditure in India? The accompanying road map expands upon this question. The legal framework required to set such a project in motion is already in place, and some aspects of government spending, such as subsidy transfers and welfare payments, are also moving towards an electronic disbursement model. However, the processes by which the Centre disburses funds to the states—from obtaining a grant under the budget to finally utilizing funds for various purposes—continue to remain cumbersome, opaque and riddled with delays, thanks to the hierarchical structure of fund flows and the lack of computerization. Money that is not delivered to the intended recipients in a timely fashion floats within the system, gathering dust in government bank accounts instead of being put to good use.

Once the EIN is implemented, information from the last mile—details of how government money is actually being used on the ground—can be uploaded by states and government departments at their own pace. This asynchronous design will allow a complete picture of outcomes in the field to be built up over time. The hub-and-spoke model we propose combines transparency with accountability and makes expenditure information freely available, opening up endless possibilities for utilizing this data. We can build our own transparency portals to track local spending; both the government and interested citizens can combine available data to come up with new and interesting insights, and third-party developers can create apps that leverage this information to provide a better customer experience. Thanks to the massive savings and improvements in efficiency that the creation of an EIN can bring about—to the order of several billion rupees—the government can deliver a vastly improved level of service to its citizens. Projects can be funded based on their outcomes, with high performance receiving its due reward. Real-time monitoring of expenses will be possible at the grassroots level, allowing the flow of funds to be better targeted. The wealth of data this system generates can be analysed to determine geographic, demographic and temporal trends, information which can be used to drive policy-making at the national level.

With better planning, better analysis and better accountability, the EIN will drive the transformation of government into an entity that

ENABLING FRAMEWORK:

- The Right to Information Act
- Electronic Service Delivery Bill
- The Information Technology Act

THE **EXPENDITURE** ROAD MAP

THINGS ARE **CHANGING:**

1. Computerization of benefits and delivery to bank accounts

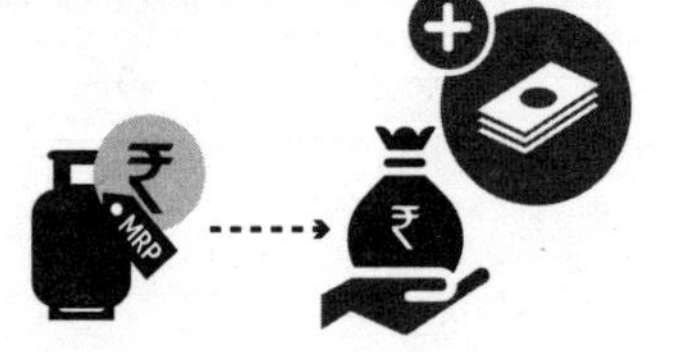

2. Direct cash transfer for subsidies

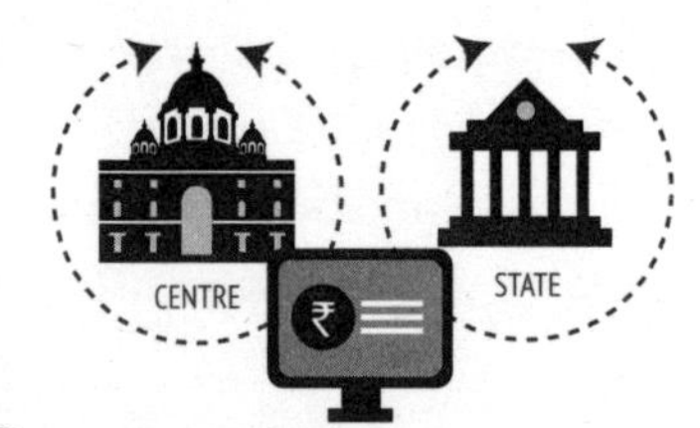

3. Computerization of government fund flow

THE **BIG** PROBLEMS:

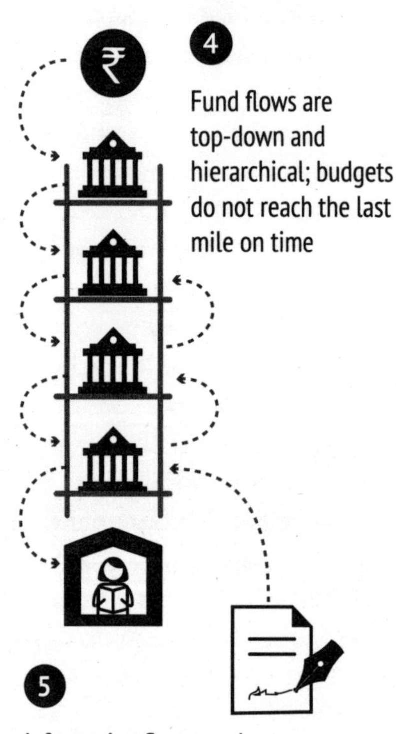

4. Fund flows are top-down and hierarchical; budgets do not reach the last mile on time

5. Information flows are bottom-up and largely non-digital

6. Funds are not distributed efficiently across government; a Just-in-Time fund flow could save Rs 1000 crore in float

TOWARDS A SOLUTION:

7 A hub-and-spoke system to integrate fund flows and information

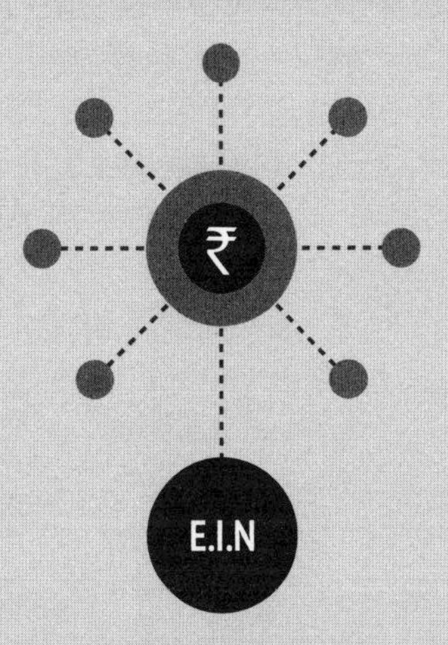

8 A transparency portal for government and its citizens

9 A professional institution to manage government expenditure

10 Greater savings through transparency and efficiency

11 Outcome-based funding, better budgeting and data analysis

is truly transparent, and one that uses its money wisely for the greatest good of the people it is meant to serve. Borrowing from Yamini Aiyar and T.R. Raghunandan, we can say, 'Money must reach where it needs to and in real time. This is the minimum of maximum governance.'

9

STRENGTHENING DEMOCRACY WITH TECHNOLOGY

> The right of voting for representatives is the primary right by which other rights are protected. To take away this right is to reduce a man to slavery.
>
> —Thomas Paine, 'Dissertation on First Principles of Government' (1795)

IN THE 2012 Manipur assembly polls, eighty-year-old Tongsei Haokip and his wife Veinem cast their votes in Saitu constituency. Not particularly remarkable, except for the fact that the couple had been dead for some time. Apparently, they were not the only ones whose enthusiasm for exercising their franchise went beyond the grave; three other dead people also voted in the same constituency.[1] On the other hand, many residents of Mumbai who happened to be very much alive found that they could not vote in the 2014 assembly elections because their names had been added to the list of deceased voters. Messages circulated on social media advising voters on what to do if they found themselves in this peculiar situation—they had to sign an affidavit that they were, contrary to the Election Commission's belief, alive and residing in the area, and hence should be allowed to vote. Also in Mumbai, Deepak Parekh, the chairman of one of India's largest banks, HDFC, found his name missing at the polling booth where he routinely voted in previous elections.[2]

The Indian public has perhaps become inured to these and other similar stories that routinely pop up in the headlines whenever it's election season, but they are indicative of a very serious problem indeed. We are the world's largest democracy, sustained by an electoral process that is a singular constant in a nation that has undergone all manner of change post-Independence. The right to vote is the one right that India bestows generously upon all its citizens, giving them the power to hold their leaders accountable. Errors in electoral rolls are in effect cheating the people out of an opportunity to influence the future of the nation, preventing them from being true participants in the democratic process. Cleaning up our electoral rolls is the first priority when it comes to electoral reform. In this chapter, we explore how technology can be used to transform India's electoral process.

A festival of democracy: India's early elections

The independent, newly democratic India that emerged dazed and bloodied from the brutality of Partition did not inspire its observers with confidence. Former British territories and quasi-independent princely states had been cobbled together into a loose alliance held together more by a strong political will rather than a fiercely cherished national identity, and the entire Frankensteinian creation was liable to start fraying at the seams in short order—a 'geographical expression, no more a single country than the Equator', as Winston Churchill put it.

And yet, this untested and unlikely coalition now ranks among the world's most stable democracies, even as other countries have managed to sustain little more than a fitful flirtation with the idea. We as a nation have displayed a remarkable dedication to the idea of free and fair elections as the lifeblood of a democratic society. The cynics among us may point to the dramatic headlines of booth capturing and rigging that pop up during every election, or feel that their vote has no impact on the political establishment. But the fact remains that every few years, whether old or young, rich or poor, farmer or movie star,

Indians faithfully line up outside their local polling booths to exercise their inalienable right as citizens of this country, that of casting their ballot to choose their future leaders.

Independence-era India was riven by a multitude of socio-economic fault lines—the average Indian at that time could expect to live only to thirty-two, and less than a fifth of the population could read and write.[3] When it came to the electoral sphere, however, all diffidence was shed for a bold, hard-charging approach where every eligible Indian was granted the right to vote, with no barriers of class, caste or gender being considered. The western world had approached universal suffrage with great caution, doling out votes first only to men, and amongst them the landed gentry. It took years of protests and civil movements before women, for example, were allowed to participate in elections. New Zealand was the world's first country to grant its citizens universal suffrage in 1893; the United Kingdom followed thirty-five years later, and it took the United States seventy-two years to catch up.[4]

In contrast, India bypassed the entire argument by granting universal suffrage from the get-go. Since then, elections in India have been a spectacle on a grand scale, a festival of democracy that the whole world looks to. While we analyse the ways in which technology can help to strengthen the voting process, it is instructional to look back and see how elections in India have evolved over the last sixty-odd years, a time in which sixteen general elections have been successfully held.

In his book *India after Gandhi*, the eminent historian Ramachandra Guha describes what it took for India to hold its first-ever general election in 1952, a process he refers to as 'the biggest gamble in history'.[5] A hundred and seventy-six million Indians were eligible to vote, and their names and other details had to be painstakingly collected and recorded to build India's first-ever voter roll. Sixteen thousand and five hundred clerks were appointed on six-month contracts to type and collate the rolls by constituency; the printing of these rolls consumed over a hundred million sheets of paper. In order to ensure that the elections were conducted smoothly, 56,000 presiding officers were employed, as were 280,000 helpers and 224,000 policemen; 389,816 phials of indelible ink were used to mark the fingers of those who had voted, and the entire exercise cost the exchequer over $1.5M (Rs 100 million).[6]

Then, as now, the press coverage of the election carried stories of villagers who walked for miles to vote, and tribals who made the arduous trek through impenetrable forests to reach the nearest polling booth. The elderly turned out too, leaning on crutches or propped up by their relatives. In a trend that persists to the current day, rural and working-class India participated in far greater numbers than the affluent middle class;[7] the *Times of India* reported that in Bombay (present-day Mumbai), long queues formed outside polling stations in the industrial areas of the city, with workers braving the morning chill, while in the fashionable Malabar Hill neighbourhood, 'it appeared as if people straggled in for a game of tennis or bridge and only incidentally to vote'.

Election officers too went the extra mile, rather literally in some cases, to ensure that the elections were conducted smoothly. One official walked for six days to attend a preparatory workshop, while a fellow colleague had to endure a four-day mule ride. Bridges were built across rivers and naval vessels were pressed into service to make sure that every remote village and island received a copy of its electoral roll. In one village, the polling station was a schoolhouse with only one door. The rules mandated that the same point could not be used for both entry and exit; the solution was to convert one of the windows into a second door, complete with makeshift steps to help voters hop out once they'd cast their ballot.

The job of managing this mammoth exercise fell to the Election Commission of India (EC), established on 25 January 1950 as an autonomous body responsible for all electoral processes in the country. Since then, the Election Commission has undergone its own share of reform in the interests of eliminating electoral fraud and ensuring that a citizen's right to vote is safeguarded.

The Indian bulldog and the year elections changed

Pudgy and dour, T.N. Seshan hardly seemed like the kind of swashbuckling hero who could take on the establishment and win.

And yet, the reforms that Seshan instituted during his tenure as the Chief Election Commissioner (CEC) completely changed the face of Indian elections. Faced with complaints that elections had now become rather dull and lacklustre affairs, Seshan acerbically retorted, 'If they want sound and colour, the cinema halls are always open.'

The organization that Seshan headed, the EC, is responsible for conducting elections in a manner both fair and free. From the day the candidates file their nominations until the results are declared, they must operate under an electoral code of conduct, and it is the EC's job to ensure that this code is strictly adhered to. The EC functions under a set of constitutional safeguards designed to protect its neutrality and independence; for example, its funding is drawn directly from the Consolidated Fund of India, the government's exchequer, without requiring approvals from any other government agencies, bureaucrats or politicians. It is a small organization by design, with only around 300 employees; come election season, that number swells to nearly 5 million polling personnel and civil police forces, drawn from the civil services on deputation. This temporary workforce dissolves once the elections are over.[8]

Despite these regulations, Indian elections were notoriously corrupt and ridden with fraud, a state of affairs that took a turn for the better when Seshan assumed office. Taking refuge in the letter of the law, Seshan in his capacity as CEC insisted upon a strict implementation of the model code of conduct. He confronted the political establishment head-on, asking the prime minister to expel senior Cabinet ministers for unduly influencing voters and threatening any politico who broke the rules with immediate disqualification. 'Al-Seshan' was just one of the many sobriquets the belligerent Seshan earned from the press; far from being embarrassed, he revelled in the image of himself as a ferocious watchdog guarding the elections, pointing out that perhaps it would be better to call him a bulldog since the resemblance was greater.[9]

Under Seshan's watch, security levels were beefed up; in the 1995 assembly polls in Bihar, a state notorious for brazen booth capturing and widespread electoral fraud, 650 companies of paramilitary forces were brought in to maintain law and order. Forced to toe the EC line, politicians now had to wind up their speeches before the official deadline. The endless motorcades, loud and colourful political rallies, the open distribution of alcohol, money and other freebies to entice voters, the many instances of booth capturing and voter intimidation—all of these became a thing of the past as the model code of conduct was rigorously enforced.[10]

Voter ID cards became mandatory, and under Seshan's stewardship a nationwide programme was implemented to issue these cards to all eligible voters. Today, these cards have become an important form of ID in India, not just for elections but for many other public and private services as well. Thanks to Seshan's aggressive tactics, elections in India have become far less of a tamasha. What they have lost in pomp and extravagance, they have gained in guaranteeing that every eligible citizen of India can cast his or her vote without coercion, bribery or fraud.

The death of paper, the birth of the machine

The electoral reforms of the 1990s set the ball rolling for a much-needed clean-up of elections in India, but their domain ended the moment the voter stepped into the booth. The actual process of casting one's vote was equally in need of an overhaul, a problem magnified by the scale on which Indian elections traditionally operate. In the 2014 general election, the number of people that turned out to vote was greater than the entire population of Europe.[11] In 1952, a staggering 85 per cent of the electorate was unable to either read or write; although those numbers have improved, a significant chunk of the Indian populace continues to remain largely illiterate. How, then, does one design, build and operate a voting system that makes allowances for this fact while still granting the right to exercise one's franchise in a free, fair and informed manner?

The early answer to this question was the paper ballot, a method that continued to remain in use until fairly recently. Simple to use and easy to set up, it required voters to stamp the ballot next to the name and party symbol of their chosen candidate before folding the ballot in half and dropping it into the ballot box. However, the system had its share of drawbacks. If the ballot was incorrectly folded, the stamped ink was liable to smear, sometimes rendering the ballot invalid. Since the paper wasn't reusable, a fresh batch of ballots had to be printed for every election, driving up costs. Tallying the stamped ballots was slow, cumbersome and prone to error. Most importantly, the system was vulnerable to fraud. Supporters of a political party could, and sometimes did, capture a polling booth by force, stuffing ballot boxes with thousands of pre-stamped ballot papers to get their candidate to win, while preventing legitimate voters from casting their vote.

Starting in 1999, the EC decided to tackle the weaknesses of the paper-voting system by abandoning it altogether in favour of electronic voting, a process chronicled in *Imagining India.*[12] As Nandan wrote, 'India's elections have typically been corrupt and chaotic, with "ballot-box stuffing" part of the nuts and bolts of getting yourself elected, and voting fraud in some areas has been as high as 40 per cent. The electronic voting machines (EVMs) considerably reduced the problem of ballot stuffing.' This 'electronification of India's elections' was the first major application of technology towards improving the voter experience. Other countries continue to struggle with electronic voting—Germany, for example—and some, like Ireland and the Netherlands, have reversed their decision on electronic voting. However, recognizing the value of such a system in cleaning up India's electoral process, the EC implemented electronic voting across the nation.

The EVMs have been designed and built indigenously to specifications laid down by the EC, with features meant to prevent fraud while also being simple enough to meet the needs of a socially and geographically diverse country.[13] They are entirely battery-powered and do not need any external power source, important in a country where the uninterrupted supply of electricity is far from a given. The physical design is rugged and compact so that they can be easily transported to the thousands of polling booths across challenging terrain. They are cheap to make and easy to use. Much like the paper-ballot system, EVMs have also been designed keeping in mind that many users are largely uneducated and may not be comfortable with operating electronic devices. The interface between the user and the device is as simple and uncluttered as possible so that the voters are not confused or intimidated into making an error. All the voter has to do is to press a button on the EVM corresponding to their choice of candidate, and their vote is recorded.

In practical terms, the EVM functions like a computer. The machines run on software hardwired into the system as they are manufactured, and any attempt to alter the software by opening up the machine causes it to shut down automatically. An EVM is also meant to be a closed system—think of a computer that's not linked to the internet or any other network but functions in complete isolation. When it's time for the machines to be used, officials from different political parties verify that they have been set up correctly and the number of votes counted for each party is zero. After the voting is complete, the machines are sealed and physically transported for storage. During the counting process, the data from each machine is collected and compiled, accounting for every machine that is deployed. The machines can then be reset and are ready for the next round of elections.

The use of EVMs on a countrywide scale has significantly improved the transparency of the electoral process. The voting process and the counting of votes have now become much faster while simultaneously eliminating counting errors due to human fallibility. It has also become difficult to capture booths and cast fake votes. The EVM allows only one vote per voter ID, and no more than five votes can be cast in one minute. Tampering with machines to register fraudulent votes takes much longer and is far more complicated than stuffing ballot boxes, and electoral fraud has dropped in consequence. Thanks to the widespread use of EVMs in state and central elections, expenditure has dropped as well. The initial cost of buying and setting up the machine is offset by the fact that it can be reused multiple times. Consider that an estimated 30–40 per cent of the total cost of an election went towards the printing of paper ballots, an expense that can be entirely done away with a one-time investment in EVMs.[14]

At present, trying to access the software that powers an EVM will probably land you in jail. But in other parts of the world, technologists are pushing for a model they have termed 'open source voting', where every EVM's software programmes would be made completely visible to the public. While editing the code would still be out of bounds, anyone with sufficient technical knowledge will be able to examine it to make sure that the EVM is registering and counting votes correctly.[15]

Any discussion of electoral reform is incomplete without the fact that India is one of the few countries in the world to have an established 'none of the above' (NOTA) ballot option, through which a voter can reject all candidates standing for election. In 2013, the Supreme Court ruled that this option must be offered during an election, writing in their judgement that this 'would lead to a systemic change in polls and political parties will be forced to project clean candidates' and increase voter empowerment.[16] In the 2013 assembly elections, NOTA was offered for the first time as a valid choice for voters, and was seen as a valuable option for those who felt that none of the candidates were worthy of their support.

Plugging holes, upgrading the machine

While the financial and operational benefits of EVMs are undeniable, they do come with their own set of concerns, particularly when it comes to data security and tampering. To overcome these issues, the EC has introduced a new system, the Vote Verifier Paper Audit Trail (VVPAT), which gives every voter a paper receipt verifying that their vote was cast correctly; such receipts can be used to check for electoral fraud as well as to audit the results from EVMs.[17] Field trials of the paper audit system have been conducted in 2011 as well as in the more recent assembly elections in Nagaland.[18] The Supreme Court directed the EC to formally introduce paper audits in a phased manner so that they could be used in the 2014 general election, and field trials included the Bangalore South constituency (from where Nandan contested the elections) as well.[19]

Another downside of EVMs in India has been that the results from every machine are disclosed during the counting process. Why is this an issue? Consider that every polling station typically has one EVM which registers about a thousand votes. Providing voting data at this level of detail makes it easy for political parties to track voting patterns in small communities and villages. Unfortunately, politicians use this fact to threaten the electorate with dire consequences if the outcome is not in their favour, diluting the secrecy of the voting process and endangering the freedom which is essential for an unbiased election.

As a result, the EC has developed a device called the 'totalizer', which delivers results at the constituency level. All EVMs used in an election, whether it's for a parliamentary constituency, an assembly constituency or a local body, are plugged into the totalizer, which collates the data and provides a final count for all candidates; it is not possible to obtain booth-level data once the totalizer has been employed. The use of the totalizer thus safeguards the identities of communities and their voting patterns, removing the fear of politically motivated reprisals. The first field trial of the totalizer was carried out on a pilot basis in the 2009 Bhadohi assembly by-poll in Uttar Pradesh.[20] With pressure from the Supreme Court, it is likely that the totalizer will become part and parcel of the Indian election process, just like the EVM, protecting the privacy of voters and ensuring that elections remain fair and free.

Resurrecting the dead and other ways to cheat voters

While stories about dead people lining up to vote might make one wonder whether the elections had somehow gotten bizarrely entangled with an episode of *The Walking Dead*, the reality is far more prosaic—the error-riddled electoral rolls allow for voter fraud to be perpetuated by simply appropriating the name of a dead person who hasn't been struck off the list. A revision of the voter rolls in New Delhi revealed that over 80,000 deceased individuals were still registered to vote.[21] This, despite the existence of procedures to officially delete a registered voter's name from the rolls after their demise. As in the case of the Haokips, whom we mentioned at the beginning of the chapter, these errors are exploited to cast fraudulent votes. Often, the names of non-existent individuals are deliberately added to the rolls, and these 'bogus voters' are another class of imposters who subvert the electoral process through fake voting. While the number of invalid votes has decreased considerably since the introduction of EVMs, an estimated 85 million names on India's voter rolls are either duplicates or fakes.[22]

The opposite problem is also true—plenty of people who are actually eligible to vote find that their names have mysteriously vanished from the rolls. In Pune, residents of eighty-eight apartments in a single housing society found that not one of them had been included in the voter list for the 2014 election.[23] And in a delicious irony, the CEC of India, Navin Chawla, almost didn't vote in the 2009 election because a clerical error meant that he landed up at the wrong polling booth, where his name was not on the voter roll.[24]

Despite the best efforts of the EC, it is not always smooth sailing for new voters who wish to register. It takes months for the registration process to be completed, in part because the process is split between two organizations—the local municipal body registers voters, but the addition of their names to the final voter roll is the responsibility of the EC's state office. The lack of coordination between these two organizations means that it is entirely possible for applications to hang in interminable limbo. In Karnataka, as in many other states, the EC, in an attempt to streamline the process, has set up a website that allows for online enrolment. Using the website to enrol as a new voter is a simple and seamless experience—until you get to the end and are unexpectedly informed that you need to go to your local municipal office to have your application verified. You're now dumped from the digital into the real world, having to go through the tedium of spending hours at the municipal office in an attempt to complete the verification process.

During Nandan's 2014 Lok Sabha campaign from Bangalore South, many of us went from door to door in the constituency, urging people to register as voters if they hadn't done so already. Many of our volunteers heard complaints from people who had used the

EC's website to register but had not received a voter ID. The EC officials could not help, saying it was an administrative issue that only the BBMP, Bangalore's municipal body, could resolve; the BBMP, in turn, expressed deep scepticism about enrolments done using 'some website' which they felt was not properly integrated with their own enrolment platform. The lack of communication between the two bodies left prospective voters in the lurch, unclear about what they were supposed to do next; in many cases, they were unable to vote in the election.

There were many other reasons that people were unable to vote. The most common reason was that after moving to Bengaluru from another city or state, they had applied to have their address updated on the voter rolls, but were still waiting for the change to be made. When urban Indians routinely hop from city to city while changing jobs, when migrants from rural areas pour into our cities to better their prospects, when most public sector jobs are transferable, why are our voter IDs tethered by geography? The wearisome wait for an address to be changed results in legitimate voters being disenfranchised for no fault of theirs.

Many of these errors of inclusion and exclusion are also deliberate. Before every election, it is almost certain that one political party will cry foul, blaming the other for manipulating the voter rolls in their favour. The EC releases voter data before and after every election, and an analysis of the numbers reveals bizarre swings in the voting population—massive fluctuations in the number of voters in a five-year period, where millions of voters mysteriously enter or drop out of voter lists, or apparently move from one state to another.[25] The actual trends in population growth and movement come nowhere close to these patterns, suggesting that names are being added and deleted en masse to further political ends. Another example comes from N. Gopalaswami, former CEC of India:

> Immediately before the 2006 general elections to the Kerala Assembly, after an analysis on the above lines, the Chief Election Officer of Kerala was able to pick two Assembly constituencies—

one each in Kasargod and Palghat—that showed an abnormal increase in the number of electors. A special check ordered by the Commission, under the supervision of two senior officers, one from Karnataka and the other from Tamil Nadu, revealed large-scale duplication of names in the polling stations on the Kerala side with polling stations in Karnataka and Tamil Nadu respectively. In the Kasargod Assembly constituency, about 5000 duplicate (bogus) voters were deleted. Incidentally, that figure matched the margin of victory in the previous election![26]

An entire political economy has sprung up around influencing new voter registrations and the composition of electoral rolls, whether the election is local or national. Voter lists are manipulated because the enrolment officer wants a bribe, or has political affiliations that dictate who he will enrol—for example, by choosing to enrol those who are likely to cast a favourable vote. The same forces come into play when voter rolls are revised for every election. Further, enrolments are the responsibility of the regular administrative machinery, which is already overburdened with routine tasks. The result is a cumbersome and opaque system which offers little recourse to those who need it and appears practically insensible to the needs of the people it is meant to serve.

Despite the electoral reforms that have changed the face of Indian elections and technological advances like EVMs, we believe that there is much to be done, as we learnt from our own experiences on the ground during Nandan's Lok Sabha campaign. We discuss some of these issues below.

Cleaning the Augean stables: Fixing India's voter rolls

Discussions about electoral reform in India tend to circle around a few hot-button topics. Campaign finances are one, with most of us agreeing that the existing legislation in this area has not been sufficient to curb the financial excesses of political campaigns. Another is whether candidates with criminal records should be allowed to contest elections and hold office as members of Parliament. These are certainly important areas for any reforms to target, and stories about excessive spending by one candidate, or the criminal antecedents of another, do tend to make the headlines. But as we met and spoke with voters in Bengaluru, and as we toured the polling booths of Bangalore South on election day, we realized that it is equally important to respect and protect the right and the desire of people to participate in elections and elect their leaders. This is the one chance granted to our people to make themselves visible to the state, and to restore the lopsided power balance that exists between the government and the people it serves.

In today's India, when your bank account and your mobile phone connection can move with you from one city to another, it is only fair to expect that a voter ID should be able to do so as well, allowing you to vote no matter where you completed the registration process and where you live now. By imposing geographical constraints on the ability to vote, we are in effect denying people their fundamental right as citizens of a representative democracy. To address these issues, we propose the creation of a centralized voter management system, explained in further detail in the accompanying diagram.

Voter registration is such a cumbersome process partly because you can only enrol at one designated location—typically in the offices of the local administration, which creates an undue burden on them while also severely restricting the freedom of choice of those wanting to enrol. Compare this to the process of enrolling for Aadhaar, which was carried out by multiple agencies at multiple locations throughout the country. What if you could enrol for a voter ID or update your details—a new address or phone number—not only at the local municipal office, but also at a public sector bank, a petrol pump, a citizen service centre, the post office, or the local railway station?

A single, centralized voter enrolment system, running the same multilingual software across the country, would allow you to enrol at multiple points. The bottlenecks that exist currently would be removed and the enrolment process could easily be scaled up to cover the entire nation. In such a system, no one enrolment agency would

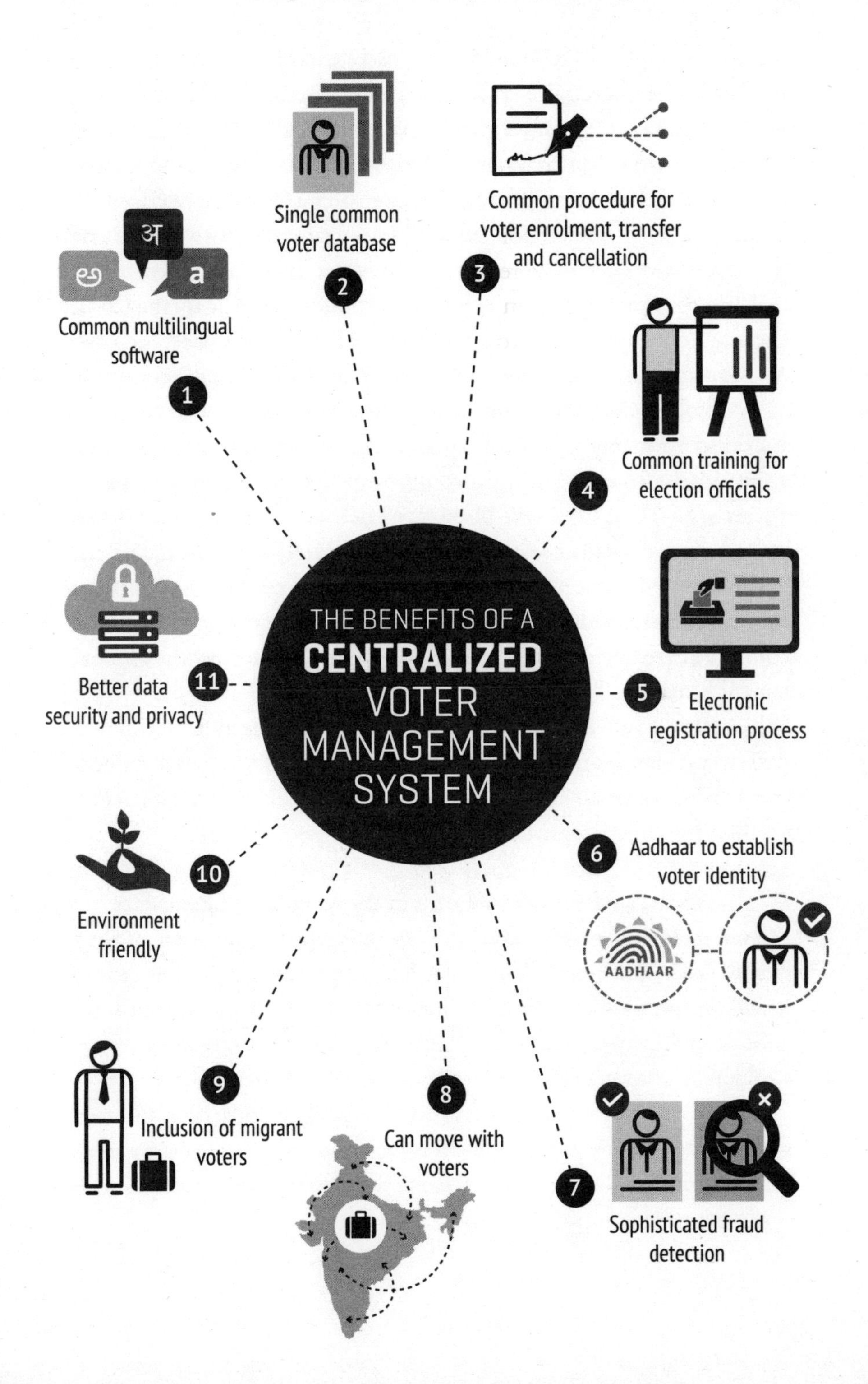
THE BENEFITS OF A
CENTRALIZED
VOTER
MANAGEMENT
SYSTEM
1
Common multilingual software
अ
a
2
Single common voter database
3
Common procedure for voter enrolment, transfer and cancellation
4
Common training for election officials
5
Electronic registration process
6
Aadhaar to establish voter identity
AADHAAR
7
Sophisticated fraud detection
8
Can move with voters
9
Inclusion of migrant voters
10
Environment friendly
11
Better data security and privacy

hold a monopoly; there would be no incentive for distortions, biases and fraud, eliminating the entire political economy that has muddied voter registrations. When we already have a centralized system capable of storing the Aadhaar data of 1.2 billion Indians, there is no reason why we can't build a similar system for voter ID data as well.

Transparency is a second, equally important attribute. As part of the government's open-data initiative, voter lists should be made available for anyone to inspect in a machine-readable format, one that any computer can recognize and process. That is not the case today; voter rolls are available only in formats that computers cannot read, which means that you have to download the data manually and build your own software to read and analyse them, unless you are up to the task of sifting through hundreds upon thousands of entries yourself. As we will discuss in the next chapter, one of the first tasks that FourthLion Technologies—a Bengaluru-based start-up that developed the technology for Nandan's Lok Sabha campaign from Bangalore South—undertook was to write the software that would digitize and clean up Bangalore South's voter lists, a process that took months.

P.G. Bhat, a retired navy commander, software engineer and social activist, has spent years scanning electoral rolls, and found over 1 million errors in Karnataka's voter rolls. An article in the *Business Standard* profiling his efforts reports:

> . . . he found a 4,818-year-old man in the voter lists of Bangalore. And there was more: Over 13,000 voters had more than one wife, 1,829 had 'female' husbands, 502 were under 18 years of age and 96 were above 120 years along with hundreds of thousands of possible voters with two—even three, records—spread across different constituencies in Karnataka. Shocked after discovering over 1 million instances of errors in the electoral rolls, Bhat took the Karnataka Election Commission to court and with some legal intervention managed to get many errors rectified before the assembly election earlier this year.[27]

What was Bhat's motivation? He is quoted in the article:

> 'I don't see what I do as activism. I am not interested in shouting slogans or gheraoing,' says the soft-spoken Bhat. He might have a different term for it, but there is no getting away from the fact that for the past couple of years Bhat has devoted most of his energy and time—up to twelve hours a day—to the one-of-its-kind cause. 'I don't want to fight the system, I want to work with it.'

While we must be grateful for the efforts of concerned citizens like Bhat, his story only highlights how desperately India's voter rolls need error correction and greater oversight, possible only by making them transparent. The Bengaluru-based NGO Janaagraha has also put in a great deal of effort into cleaning up voter rolls, working together with the EC on a system they've christened PURE (Proper Urban Electoral) Roll Management System. A survey they performed in New Delhi in 2015 found that nearly 22 per cent of the names on the voter list need to be updated or deleted as these individuals were not found at their listed address.[28]

Any changes to the voter list should be clearly visible, and all political parties and citizens should be brought on board with the process so as to forestall any allegations of malpractice. Equally important, individuals should be able to view their voter records easily so that any errors in their record can be quickly noted, a process that can be made faster through automation. Once the data is available, political campaigns can use it as a single source of truth. The data can be accessed with a simple device like a smartphone or a tablet. As we will discuss in the next chapter, one of the most successful technological innovations in Nandan's campaign was also among the simplest—a digital voter roll that people could look up on a smartphone, making it easy to find voter data and to direct people to the correct polling location. Many volunteers told us of people who landed up at the wrong polling booth, or whose data was missing in the list of voters, and who would have had to return home without casting their vote if volunteers hadn't looked up the correct data on the digital roll. Now imagine if we could build a digital voter roll for every voter in the country, one that was accessible to both officials and the people. The electoral process would become much smoother for everyone, errors would be quickly caught and corrected, and nobody would be denied their lawful right to vote.

Elections 2.0: The death of fraud

In our opinion, the most effective way to strengthen India's electoral process is by linking voter IDs to Aadhaar numbers, a view that is shared by the country's current administration. Earlier this year, former CEC H.S. Brahma announced that the databases of Aadhaar and the voter ID—the Electoral Photo ID Card, or EPIC—would be synchronized with each other, a task expected to reach completion by early 2016. In a public address, he was quoted as saying, 'We have decided to embed the Aadhaar data on our platform of the EPIC . . . Once we are able to do this, we will have 100 per cent purity of the (electoral) rolls. Once these two data come together there will be correct name, biometrics and address of an individual. The day we do that, 99 per cent of complaints of political parties and candidates about electoral rolls will disappear.'[29] The scheme to link Aadhaar with the voter ID was officially launched in March 2015 under the moniker of the National Electoral Roll Purification and Authentication Programme (NERPAP).[30]

The potential benefits of meshing Aadhaar with the voter ID rolls are explained in the accompanying diagram. The linkage will automatically confer Aadhaar's key attribute—uniqueness—to a voter ID, and the individual's biometric data can be used to authenticate their identity. Once the two are connected, cross-checking IDs and weeding out duplicates would become an easily solvable problem. It would also be remarkably effective at eradicating voter fraud, since a person with the same biometric data can't vote again or vote as someone else. Creating a fake voter profile would become so complicated that no would-be election rigger would bother trying. As Brahma commented, 'India will be the only country in the world where there will be complete biometric of voters. Also, there will not be a single fraud or duplicate voters.'[31]

Once the Aadhaar linkage is in place, a suite of 'smartphone first' applications can allow every aspect of the voting process—voter registration, address changes, polling booth information, perhaps even casting one's ballot—to be available to us on our smartphones. In a perfect world, we should be unrestricted even in the choice of polling stations—rather than having to vote at a specific polling booth to which we have been assigned, we should be able to walk into any polling station in any part of the country, have our voter ID number and identity obtained and verified through the cloud, and cast our vote instantly. Estonia, one of the most digitally advanced nations on the planet, has already leveraged its national ID programme to allow citizens to vote using their mobile phones; one happy citizen took to Twitter to share that it took just two minutes to vote—'being hundreds of miles away from home. How awesome is this?'[32]

It may be a while before India's voting process reaches equal levels of awesomeness, given that hundreds of millions of people have to cast their vote on election day and the entire process needs to be completely watertight. In the meantime, using Aadhaar for identity verification during the voter registration process would be a good start. Every enrolment or update centre designated by the EC should be equipped with a biometric reader. The registration process will use Aadhaar to verify the person's identity and demographic data. The new address details will be downloaded by the EC's system and the voter will be assigned a polling station at which they can vote. The cut-off for revisions can be set at one month before the election, at which time the final voter rolls are prepared. The delivery of the updated voter ID can be the responsibility of the postal system, in the same way that we worked with India Post for the delivery of Aadhaar letters. This can complete the final leg of address verification, while simultaneously saving the voter from having to make several trips to the point of enrolment.

One interesting use of electronic voting has recently been proposed by the government to allow military personnel to vote.

Currently, they receive ballot papers in the mail, which they have to fill in and return via India Post. Under the new proposal, once their voter IDs have been linked to Aadhaar numbers, selected areas will be equipped with biometric authentication devices; Aadhaar authentication will be used to verify their identity, after which they can cast their vote electronically. Officials envision that such an electronic voting system will act as a 'technology-driven solution where people can vote, which reaches on time, does not get lost or is not tampered with'.[33]

Former CEC N. Gopalaswami suggests that Aadhaar can also serve as a tagging device for new voters to be enrolled. Since Aadhaar numbers are now being issued to all Indian residents from birth onwards, the database can be set to automatically flag those Aadhaar holders who turn eighteen in a given year, making them eligible to vote. Rather than putting the onus of enrolment on the citizen's shoulders, the government itself can now initiate the enrolment process for such individuals. The many benefits of an Aadhaar-linked voter ID are explained in the diagram that follows.

The future of your vote

Using many of the ideas we delineate above, we can take our voter lists and elections to the next level of transparency, accountability and fairness. Technology can help to make the electoral process more convenient for citizens. A centralized voter system will make life easier for citizens while eliminating fakes, ghosts and duplicates. It will be capable of spotting fraud as it happens—a system that credit card transactions already rely on. If it can work for your credit card, it should work for your vote.

Elections are the foundation of our democracy, a democracy that represents an unflinching commitment by our country to every citizen, granting them a voice in the form of a vote. If citizens lose faith in our electoral process, we lose one of the cornerstones of our nation. A constant process of institution-building over the decades has seen us withstand many challenges. It is only technology that gives us the

Voices of migrant citizens remain unheard since franchise is tied to a single location

Millions of fake, defunct voter cards exist leading to fraudulent voting

Getting a new voter ID is tedious. An existing ID is also not a guarantee of appearing in the voter list

Vote from anywhere convenient

Authenticate identity at the time of voting

Make it easy to get a voter ID

THE BENEFITS OF GOING ELECTRONIC & LINKING TO AADHAAR

FREE AND FAIR ELECTIONS ARE THE CORNERSTONES OF DEMOCRACY

Anything that prevents people from voting or gives undue advantage to anyone works against the principle of democracy

kind of scale and reach we need to further strengthen our electoral processes and build a truly inclusive democracy, one in which every citizen can exercise their most fundamental right, that of choosing the leaders who will shape the future of their country.

10

A NEW ERA IN POLITICS: THE PARTY AS A PLATFORM

> Elections belong to the people. It's their decision. If they decide to turn their back on the fire and burn their behinds, then they will just have to sit on their blisters.
>
> —Abraham Lincoln

WHEN NANDAN DECIDED to contest the 2014 Lok Sabha elections from the Bangalore South constituency, he reached out to several people for ideas and advice on how to plan and execute his campaign. One of them was Andrew Claster, deputy chief of analytics for Barack Obama's 2012 presidential campaign. Nandan, Viral and two other colleagues, Shankar Maruwada and Naman Pugalia, sat down with Andrew to strategize. Obama's campaign had made headlines for its innovative, heavily technology-driven approach; what were the main principles on which it had been run? To our surprise, Andrew pointed us to a document that had been written 174 years ago, by none other than Abraham Lincoln.[1] In short, the entire campaign plan consisted of building a multilayered organization designed to reach out to every voter, meeting the voters face-to-face and convincing them to vote for your candidate. These fundamentals of electioneering are as relevant today as they were 174 years ago; in Andrew's words, 'Many of the innovations of today's campaigns involve simply employing technology, data and predictive modelling to enhance and make more efficient the same basic functions as were outlined by Lincoln in 1840.'

The campaign that Nandan and his team eventually ran took the standard operating manual of Indian politics—public campaigns, rallies, photo-ops, constituency tours—and garnished it with a liberal dose of technology. Whether it was creating a digital voter roll, running targeted advertisements or interacting with voters, technology was the invisible scaffold on which all the programmes were built and run. And our approach was far from being an outlier; as we shall see, many players in today's political arena are using technology to establish stronger ties with the people.

We believe that the software, technology and processes that political parties build will serve as valuable intellectual property for them, creating a platform that every candidate will be able to use. This will then become the lifeblood of the campaign and the central playbook from which all candidates can draw, ensuring the delivery of a strong and consistent message that people can relate to. Despite the huge effort involved in creating this sort of centralized model, political parties stand to benefit a great deal. In the long run, candidates will choose to stay with parties that make it easier for them to run their campaigns. It will also create a cadre of trained party workers, giving parties a competitive edge when it comes to gaining and retaining good candidates.

We've already seen this in effect in the past few elections. The Aam Aadmi Party (AAP) campaigns in Delhi in 2013 and 2015 were, to borrow software terminology, Version 1.0 of the party as a platform. Narendra Modi's national campaign in 2014 was Version 2.0, and perhaps the next general election of 2019 will be Version 3.0, the full-fledged implementation of the 'party as a platform' strategy, with nationwide scale and the ability to manage the electoral campaigns of all candidates within the party. The diagram that follows illustrates what we think the new face of technology-driven campaigns will be.

1 Connect with voters through the smartphone

2 Instant Polling to figure the right message

THE NEW TECH-ENABLED **ELECTION CAMPAIGN**

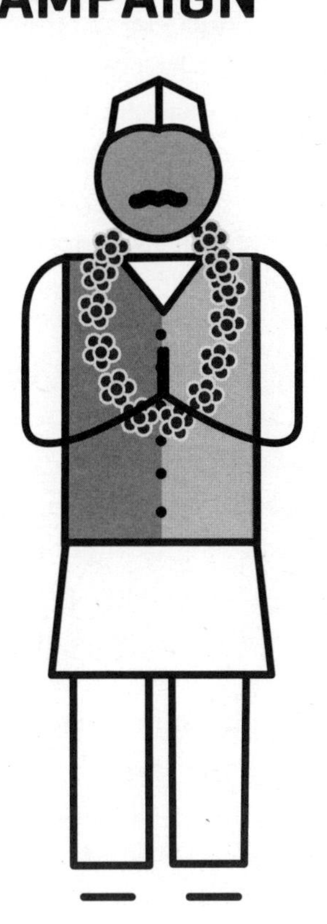

3 Effective communication strategies—the right message for the right medium

4 Enable a volunteer field force

5 Allow voters to effectively file grievances

New brooms sweep clean: The emergence of the Aam Aadmi Party

Starting in November 2013, many residents of New Delhi—housewives, rickshaw drivers and small-business owners alike—received phone calls from supporters of the AAP, asking if they would 'vote for the broom'. Not too unusual, except for the fact that the volunteers making these calls, people like Mohan Thirumalai, an IT manager, were sitting not in Delhi, or even in India, but thousands of miles away in the USA. Every night, Thirumalai would start making the first of around sixty calls, drumming up support for the party. Many other volunteers took sabbaticals from their jobs, or quit altogether and came back to India just to work for the AAP's cause.[2]

This was just one of the many innovative strategies that the AAP employed to contest elections in New Delhi. Many of these ideas arose from an earlier, non-political movement, India Against Corruption (IAC), spearheaded by the Gandhian reformer Anna Hazare; the AAP founder and maverick former bureaucrat Arvind Kejriwal was also a prominent figure in that campaign. The IAC was the first social movement in modern memory to be embraced enthusiastically by the middle class, otherwise notoriously gun-shy. How did they manage to achieve this connect with the people? They were quick to recognize and exploit the power of technology to reach out to large numbers of people while keeping costs low.[3]

One example was their use of that ingenious, uniquely Indian method of communication—the missed call. In IAC's missed call campaign, those who wished to express their support simply had to dial a specific number and then hang up; a bulk SMS campaign was launched to encourage people to call. In a span of three months, IAC racked up over 25 million missed calls. Their Facebook page had over a million followers, while their smartphone app had as many as 50,000 users.

The AAP broke away from the IAC with the goal of taking their anti-corruption agenda into the political sphere. The party's decision to contest the 2013 Delhi assembly elections pitted them in a David-versus-Goliath contest against the ruling Congress party as well as the

Bharatiya Janata Party (BJP), both of whom had decades of political experience and a well-established party structure, cadre and funding to draw upon. Instead of trying desperately to overcome their underdog status, the AAP flipped the script and embraced it enthusiastically. Part of the reason, of course, was sheer necessity—without money, big names or a strong support base among the people, trying to follow a traditional campaign philosophy would be political suicide. The campaign strategy they eventually adopted also demonstrated a great deal of innovative, street-smart thinking.

The AAP party symbol is the perfect illustration of this principle. Every political party has to choose a symbol from a pre-approved list created by the Election Commission. The AAP chose the lowly broom and turned a symbol of domestic drudgery into a potent weapon against corruption, declaring that they would clean up the Augean stables of local government. All party members wore Gandhi-esque caps emblazoned with the words, '*Main aam aadmi hoon*' (I am the common man), reinforcing the Gandhian connect that had initially begun with Anna Hazare. Poll promises were equally populist, including assurances to provide free water and to slash electricity rates.

The entire AAP campaign was largely volunteer-driven, and many of these individuals were well-educated students and professionals who came on board only because they identified so strongly with the party's anti-corruption agenda. As we've said, the middle class tend to largely stay out of the heat and dust of politics, and the fact that these volunteers were hitting the streets of Delhi in large numbers, going door-to-door and urging people to vote for the AAP, certainly created food for thought.

In parallel, the AAP also engaged heavily with social media to recruit volunteers and propagate party ideology. Kejriwal himself took to Twitter and amassed hundreds of thousands of followers in the process, while the AAP's official website also developed a huge following. Crucially, one could also become a 'digital volunteer', helping to spread the AAP's message online without actually having to knock on doors. A huge number of volunteers were recruited in this manner, making the AAP's online strategy extremely successful.[4]

The AAP also took technology to what has always been a contentious issue in elections—campaign financing. In keeping with their motto of clean governance, the AAP decided to be almost aggressively transparent in their fundraising efforts. They filled their campaign coffers entirely through donations, and claimed that every single rupee donated was logged on the party website. If you navigated to the Donations section on their home page, you could see the names and locations of donors, and the amount they chose to contribute. Even a one-rupee donation was dutifully displayed. When the party reached its campaign total of $3M (Rs 200 million), Arvind Kejriwal took to Twitter and other social media, requesting people to stop contributing further. This was a clever strategy on two fronts: it enabled the party to build a treasure chest while simultaneously emphasizing their commitment to transparency.

The AAP managed to pull off an unprecedented victory in the assembly polls, winning twenty-eight out of seventy seats, a performance which left even the most seasoned pollsters thunderstruck. It was the first warning shot across the bows of the Indian political establishment that election campaigns were entering a new era; either you kept up, or you lost.

Twenty-eight seats, however, was not a majority, and the AAP government in Delhi folded after only forty-nine days in power. When they contested the Lok Sabha election later that year, they did not win a single parliamentary seat from Delhi. Clearly, the strategy that had originally carried them into office would need a rethink if the AAP intended to remain a serious contender in the Indian political arena.

With the 2015 Delhi assembly elections around the corner, it was time for introspection to understand why the party had lost its way. Three major reasons emerged: there was a perception that AAP had 'run away' from government; some saw its approach as one of perpetual confrontation; and others found it overambitious, wanting to expand at the cost of providing good governance at the local level.

The AAP adopted several new strategies to tackle these perceptions head-on. Firstly, they apologized for resigning from government.

Next, they decided to transform from a party that focused solely on corruption to one that took a more holistic approach towards key political issues. They focused on strengthening the party organization and chose to devote their entire energy only to the Delhi campaign.

Grassroots campaigning had always been the ace up the AAP's sleeve, and they wisely chose to continue their efforts in this direction. Just as they had previously, they built up a strong volunteer base that spread out across the city, communicating the party's message; in the last few days before the elections, there was an extra push through a Buzz campaign, which extended to biker rallies, flash mobs, street plays and even a music group, to grab eyeballs. Another way to reach out to the people was through 'jan sabhas', where Arvind Kejriwal and other senior leaders would address the gathered crowd.

The AAP continued to innovate when it came to fundraising, whether it was collecting money in bedsheets during the jan sabhas or offering a chance to dine with Arvind Kejriwal for $300 (Rs 20,000). Embracing the selfie craze, they came up with a scheme where you could win a chance to pose for a 'Selfie with Mufflerman' (a tongue-in-cheek reference to Kejriwal's ubiquitous winter accessory) for $8 (Rs 500).

One of the most striking features of the revamped AAP campaign was its decision to open-source its manifesto. Manifesto creation is usually a closed-door exercise, with limited participation by the people—in fact, this was exactly how the AAP had prepared its manifesto in 2013. This time around, they picked twelve focus areas, including such issues as women's safety, urban infrastructure and rural development. Each area was then made the topic of a fortnight-long Delhi Dialogue, in which the people's opinions were solicited. The points gathered from each dialogue became the basis of the seventy-point manifesto the AAP released. As an illustration, when youth empowerment was on the agenda, volunteers first met with student representatives, NGOs and other stakeholders to hash out a list of forty to fifty specific ideas. These ideas were synthesized into five concrete promises that Kejriwal delivered to an audience of 5000 at Jantar Mantar in November 2014. The attendees were invited to provide feedback after the event, either through paper slips or a dedicated website. All of these inputs were incorporated as well before the final manifesto was readied.[5]

The AAP's efforts ultimately paid off in a huge way. The party that had failed to make any impact on the national stage came roaring back, grabbing an astonishing sixty-seven out of the seventy seats in the Delhi assembly. Their success is perhaps best described by a conversation that leading Indian journalist and former editor-in-chief of the *Indian Express* Shekhar Gupta had with a fund manager he met in a hotel elevator.[6] Discussing the prospects of various political parties, the fund manager told Gupta, 'Modern politics has now become like the IT industry. Just as smart tech start-ups keep disrupting established giants, a political start-up has disrupted established, big parties. The big question now, do they understand this? And how will they respond?'

Narendra Modi: The triumph of the 'chaiwalla'

US President Barack Obama's two presidential campaigns, in 2008 and 2012, are widely acknowledged as having written the blueprint for a new, technology-centred approach to contesting an election, relying heavily on data collection and analysis to power campaign efforts. Journalist Michael Scherer profiled the 2012 campaign for *Time* magazine, writing that 'the role of the campaign pros in Washington who make decisions on hunches and experience is rapidly dwindling, being replaced by the work of quants and computer coders who can crack massive data sets for insight', and calling the campaign the curtain-raiser for a data-driven era in politics.[7]

No doubt Obama's 2012 campaign was closely watched in India by the BJP, and particularly by its charismatic leader, Narendra Modi, who had already demonstrated his enthusiasm for technology-based outreach and solutions during his tenure as chief minister of Gujarat. In its heavy reliance on data and analytics, the 2014 campaign that eventually propelled Modi to the PM's chair borrowed a leaf out of the Obama playbook, albeit for a larger playing field than Obama ever had to contend with. In the US, the voters directly elect their president, whereas in India, votes are cast only for the local MP, with the majority party in Parliament (or in a coalition) then electing the PM. However, Modi was acknowledged as the BJP's prime ministerial candidate well before campaigning was underway and ran a strong, centrally driven national campaign for the party, an uncommon approach in India. As we will see, he bypassed traditional media channels to establish a one-to-one connect with the people through social media and similar avenues.

The key issues that Modi and the party chose to focus on were corruption, the rising prices of everyday staples and, in the larger picture, growth and economic development as an engine for driving national progress. Recognizing that the idealism of newly registered voters and the youth would make them a powerful voter base, many of the messages were targeted at them using channels like social media which had a greater reach in this segment.

Much of the messaging effort was handled by Penn Schoen Berland, a leading US-based market research and consulting firm that had previously worked on political campaigns for such internationally known leaders as Bill and Hillary Clinton and Tony Blair. Slogans like '*Ab ki baar Modi sarkar*' (This time, a Modi government) became so ubiquitous that they were the punchlines for a whole series of internet jokes and as easily recognizable as the refrains from popular advertisements. Another tagline that resonated deeply with people was '*Acche din aane wale hain*' (Good days are going to come), reflecting the people's desire for better prospects.

Every possible channel was used to bombard voters with party messages: TV, radio, print, online ads, bus shelters, banners, newspaper inserts, pamphlets and more, in multiple languages so as to reach the local populations. This carpet-bombing was actually the result of data analysis and careful planning so that the campaign got the maximum bang for its advertising buck. The effectiveness of these marketing strategies was measured by constantly monitoring national and international media, and making tweaks when needed.

Just like the AAP, the BJP also made sure to stay in the media headlines. A textbook example was when Congress leader Mani Shankar Aiyer made a comment alluding to Modi's humble origins as a '*chaiwalla*' (tea seller) at a train station. The normal response would have been for a BJP spokesperson to roundly condemn the statement, and perhaps display some theatrically ruffled feathers on the nightly news debates; instead, the remark became the genesis of the BJP's media-friendly '*Chai pe charcha*' (Discussions over tea) platform, where tea stalls across the country were outfitted with video links and internet connections so that customers could participate in the discussions. What could have been one of a series of political insults that parties trade amongst each other instead became a major embarrassment for the Congress, and a public relations coup for the BJP.

Given Modi's fervent embrace of social media as the Gujarat CM, it is no surprise that this attribute carried over to his prime ministerial campaign as well. He was well aware of the one-to-one connect that social media can establish between a leader and his people, and had exploited this power liberally to position himself as a leader uniquely connected to the pulse of the entire state he oversaw. Now, multiple teams were created to run the BJP's social media campaign. Modi continued to further strengthen his presence on networks like Twitter and Facebook, gathering millions of followers and easily becoming the most visible Indian politician in this space, with over 5 million Twitter followers and nearly 20 million Facebook likes as of today; the *New York Times* called him a 'juggernaut of political social media'.[8] Among public figures with political leanings, he trails only Barack Obama, the Dalai Lama and Pope Francis in popularity.

In parallel, the BJP teams also recruited digital volunteers, much like the AAP did in Delhi, with the goal of amplifying their message across the online world. India272+ (referring to the number of seats the BJP would need to win to establish an absolute majority in the Lok Sabha) was a website set up to enrol volunteers and coordinate their activities, and also functioned as a major channel for message propagation. In addition to the India272+ website, the campaign also used social media to recruit and coordinate a final tally of over 5 million volunteers; the 2012 Obama campaign, by comparison, had just over 2 million.

The cornerstone of any and all political campaigns in India, no matter how large or sophisticated, continues to remain the political rally. The image of a prospective leader, hands folded as they greet the sea of people who have been waiting patiently to hear them speak, remains the quintessential picture of an Indian election. Modi's team used technology to tweak this campaign workhorse, scanning the social media air waves and drawing upon volunteers to gather local intelligence. As a result, every speech had a strong local message, and was often garnished with a smattering of words in the local language.

Each rally was converted into a mini media event by the BJP's IT cell. The speeches were live-streamed on multiple video sites, including the India272+ website. The rallies were also captured on video and handed over directly to TV channels to ensure maximum coverage. Photographs were reposted on social media, and the feedback between the party's own promotional efforts and the media coverage of each event resulted in a self-perpetuating loop of favourable publicity.

In states like Uttar Pradesh and Bihar, the BJP's public profile required a heavy boost. Here, traditional outreach methods were supplemented with such innovations as traveling 'raths' (chariots) equipped with LED screens broadcasting a pre-recorded speech by Modi. Party workers accompanying the raths handed out promotional material like caps and masks while trying to recruit new volunteers. Recognizing that political rallies are not that different in spirit from rock concerts (the charismatic lead attraction, screaming audiences and, of course, the interminable wait for the main act to hit the stage), the BJP also staged 3D hologram rallies (a favourite of the erstwhile 'King of Pop' Michael Jackson). Equipment would be hauled on trucks to the remotest parts of the country, and a lifelike hologram of Modi would deliver a pre-recorded speech in areas that he was unable to reach physically.

The success of all these efforts can be gauged by the fact that the BJP, with Modi at the helm, romped home to victory in the 2014 elections, capturing an absolute majority with ease. It was technology that made the BJP campaign a well-oiled machine, and will now be the de facto standard for the political campaigns of the future.

From technocrat to politician: Nandan's campaign from Bangalore South

Political will is essential to spearhead change at the national level, and so it was that Nandan threw his hat into the political ring, deciding to contest the 2014 parliamentary elections and eventually standing as the Congress party candidate from the Bangalore South constituency. In his own words, 'Part of the reason I stood for election was my frustration at my inability (in five years at the UIDAI) to convince politicians of the possibility of solving India's challenges with new and innovative methods.'

He started building a technology team to develop the software that would power his campaign by reaching out to three former colleagues, Shankar Maruwada, Viral and Naman Pugalia, each of whom brought complementary skill sets to the table. Viral had a strong background in technology and public policy. Naman, a former member of Google's public policy team, had helped to conduct Narendra Modi's much-publicized Google hangout, and had a keen understanding of how both traditional and online media channels could be leveraged. Shankar had years of experience in marketing and analytics. Together, the trio went on to create a company called FourthLion Technologies—a reference to India's national emblem, the Lion Capital of Ashoka at Sarnath, where the fourth lion that is hidden from the viewer's gaze was envisioned by the trio as the citizen, the invisible pillar of any democracy.

The first step: The voter list

The first order of business was to study the turf on which Nandan would be conducting his political battle. The Bangalore South constituency is made up of eight assembly constituencies (Govindraj Nagar, Vijayanagar, Chickpet, BTM Layout, Bommanahalli, Jayanagar, Padmanabhanagar, Basavanagudi), each represented by an MLA, and further divided into sixty-one wards, each of which is represented by a corporator. The eight assembly constituencies (ACs) each have their own unique character and flavour. Basavanagudi, one of the oldest parts of Bangalore, dates back to the 1500s, and still retains some of the old-fashioned culture of a bygone age. At the time it was built, Jayanagar was Asia's largest planned locality; its tree-lined streets and busy marketplaces make it one of the city's more desirable addresses. The narrow lanes and bustling shopfronts of Chickpet house traders from all parts of the country. Bommanahalli is a place of extremes—swanky glass tech parks on one hand, and a large number of garment factories on the other. BTM Layout echoes this dichotomy, housing both sprawling apartment complexes with the latest amenities and some of the largest slums in the city. Clearly, building a comprehensive picture of Bangalore South would necessitate understanding and connecting with the many identities that make up the constituency.

Once these constituency-mapping efforts were underway, we were faced with the next big question: exactly how many voters did Bangalore South harbour? At the start of election season, there were 1.8 million registered voters in the area, a number which swelled to 2 million on election day; no doubt that increase was due at least in part to our strenuous efforts at getting people registered to vote. Bommanahalli had the largest number of voters, around 350,000, while the other constituencies had close to 200,000 voters each. Going down a level, each ward had about 10,000 households, and a rough estimate of three voters per household meant that every ward had roughly 30,000 voters. None of our technology-based approaches for voter outreach would work unless we could build an electronic database housing the voter records for all of Bangalore South, basically recreating the EC's electoral roll in digital format.

In their entirety, the voter rolls represented 80,000 pages of information that we needed to digitize, with around thirty names on each page. At the risk of stating the obvious, this was a Herculean task, and it took us months to build a digital voter roll whose data quality we found satisfactory. Providing accurate digitized voter rolls that all party workers can access through a smartphone is going to be the bedrock of the party as a platform.

The people behind the campaign

Once Nandan's campaign began in earnest, it could be divided into three phases. The first was the awareness phase, where more people were made aware of Nandan's achievements and political ideology. During this time, we also ran a voter registration drive. The second was the persuasion phase, which we spent trying to convince people to vote for him, and the third was the turnout phase, where we had to make sure that people who supported him actually made it to the polling booth to cast their vote. We concentrated exclusively on the issues that impacted the daily lives of the voters of Bangalore South; in a phone call with Jack Markell, the governor of the US state of Delaware, he reminded us of that oft-quoted political aphorism, 'All politics is local.'

One of the lessons to be learnt from recent political campaigns is the need for a formal, well-organized campaign structure. Usually, campaigns are divided among four major departments: communications, legal and compliance, research, and field operations. The work of these departments is bolstered by technology, data and analytics.

What do each of these teams do? The communications team designs the initial campaign message and broadcasts it through various communication channels, testing its effectiveness and making refinements. In the persuasion phase, the team focuses on spreading the message through all the channels it has access to—print, media, social media, banners, posters, leaflets, billboards and the like. In Nandan's campaign, Devi Pabreja led this effort; she did much of the research for Nandan's earlier book *Imagining India* and played a key role in the early days of Aadhaar. The legal and compliance team, which ensures that the campaign adheres to the Election Commission's code of conduct, was anchored by Deepika Mogilishetty, a former UIDAI colleague who had worked on the legal aspects of the Aadhaar programme.

The research team handles all the polling and surveys required during campaigning. Polls are essential for the candidate to test messages for persuasion, to detect trends in public opinion, and to measure the effectiveness of the campaign over time. A good research team should be able to accurately predict the outcome of an election well before a single vote is cast. All our work on polling, data gathering and analytics was anchored by Dr Venkata Pingali, holder of a PhD in networking from the University of Southern California, and an entrepreneur focusing on energy analytics.

The field team is often the largest team in a campaign, and is responsible for door-to-door campaigning. Ideally, a field team needs one volunteer for every 100 voters. In reality, that number is closer to one for every 500. Given that the average parliamentary constituency has 2 million voters, a campaign needs a volunteer force of thousands to make contact with individual voters; it is said that the campaign message must be delivered to every voter at least five times in a successful campaign. Our field team was led by Nita Tyagi, a former entrepreneur and consultant with a passion for politics.

A profile for Mr Nilekani

If you want to persuade someone to vote for you, they need to know who you are, which political party you are affiliated with and what your ideology and campaign platform are all about. This kind of awareness-building is easier if the candidate already has a fairly high local profile. In Nandan's case, however, we had the unique challenge of a candidate who was well known both nationally and internationally, but a relative cipher to the voters of South Bangalore. Initial testing revealed that people across socio-economic groups had heard of Infosys and Aadhaar, but had no idea that Nandan was closely involved with either; many of them didn't even recognize his name.

As a candidate, Nandan always strove to build a one-on-one connect with each voter. How could we maintain the intimacy of such a dialogue on a constituency-wide scale? The campaign did several things to introduce Nandan to the voters of Bangalore South, including messaging through traditional media, social media, and personalized letters. Each letter greeted the resident by name before introducing Nandan, listing his past achievements and explaining his vision for building a better Bangalore. A small tear-away section at the bottom of the letter contained such useful information as the location of the individual's polling station and their serial number on the voter list. We heard back from a number of people who were deeply impressed at receiving a personal letter from a candidate.

We conducted polls to measure the effectiveness of all these strategies at boosting awareness of Nandan as a candidate. Starting at an unhappy low of 25 per cent, we saw awareness registering a massive jump over time, reaching around 80 per cent by the time election day came around. It was technology and data analytics that powered this leap, allowing us to choose the media and the message likely to have the greatest impact and then monitoring the results in real time.

The campaign rolls out

Once people knew who Nandan was, what he had achieved so far and what he hoped to do for Bangalore South, we now had to tackle the challenging task of persuading people to vote for him. Volunteers called the residents of Bangalore South through call centres, familiarizing them with Nandan's biography and emphasizing the local connect, starting with the fact that he was born in Bangalore South. In the later stages of the campaign, the live calls were complemented by automated phone calls that played a pre-recorded message from Nandan urging people to cast their vote in his favour. We ran polls alongside to measure how effective these strategies were.

Party workers from the Congress swept through neighbourhoods, visiting as many houses as possible, talking to the residents about Nandan's candidacy, asking them what issues they faced in their locality, and collecting their phone numbers in case they wished to be contacted further. The same task also fell to the lot of the volunteer group that emerged to support Nandan's cause, who called themselves 'Together with Nandan', or TWN. The TWN volunteer army had over 1500 foot soldiers, with 300 people fanning out into every nook and cranny of Bangalore South in the last month before the election. They were an eclectic lot—some of them were college students, others were former colleagues from Infosys or the UIDAI. Saroja Yeramilli, the TWN coordinator, quit her position at Dell to join the campaign; Meeta Karanth postponed her start date at McKinsey Consulting. We even had celebrity endorsements from such well-known Kannadigas as the actor, director and playwright Girish Karnad.

None of the TWN members were politically motivated; all of them had joined the effort only because they supported Nandan's candidacy, and worked without any pay for their efforts. In fact, many of them had no background in election campaigning whatsoever; one star TWN field worker, Kavitha, was a housewife who always took her little daughter Sahana along when knocking on doors, reasoning correctly that people would be more willing to talk to a lady with a young child perched on her hip. Sahana eventually became the TWN mascot, and could reel off the entire 'sales pitch' flawlessly, much to everyone's amusement, admonishing people in her sweet three-year-old lisp to 'Vote for Nandan, okay?' TWN members coordinated their activities using social media, especially WhatsApp, and used mobile apps to record voter interactions and maintain records of all the houses they visited. Some of the TWN members also came up with fresh ideas to persuade voters, like the flash mob that danced to popular film music at the Royal Meenakshi Mall in South Bangalore while wearing TWN T-shirts, finishing up their performance by exhorting the curious onlookers to vote for Nandan.

As Congress party workers and TWN volunteers conducted an extensive field campaign, Nandan too embarked on a punishing schedule of public meetings and padayatras, touring various parts of the constituency, meeting the local leaders as well as the people of the area. We gathered data from each of these activities to check their effectiveness, and ran polls to continuously gauge the health of the campaign. We also reposted photos of these events on Nandan's Facebook and Twitter pages, further amplifying the media coverage.

Getting out the vote

The final stage of a campaign is the turnout phase, which begins two days before voting day. At this point, all campaigning stops, and the entire effort now boils down to ensuring that supporters come to the polling stations to cast their votes. Once Nandan's campaign reached this phase, the technology team built an extremely simple voter look-up tool. This tool was nothing more than the digital electoral roll in a searchable format; entering a person's name or voter ID number into a bare-bones, text-based search box accessible over the internet drew up their entire voter record, including their serial number and the name and address of the polling station where they were supposed to vote.

On election day, political parties usually set up tables outside each polling booth with a copy of the voter roll. Party workers help voters to find their name and serial number on the list so that they can vote. Given the number of errors that the voter roll is riddled with, it isn't surprising that many people can't find their names, and have no idea whether they are even at the correct polling booth.

The day Bangalore went to the polls, 17 April 2015, we sent TWN volunteers to polling booths across the constituency, hoping to assist party workers in looking up voters. The data collected during the awareness and persuasion phases of the campaign revealed which locations needed a little extra push in terms of getting voters to the booths on D-day, and that's where our volunteers went. Regunath, principal architect at Flipkart and a TWN volunteer, told us what happened next:

> I got to the booth, and there was this gentleman standing at one of the tables looking really upset. The party worker was trying to find his name in the voter rolls, but it wasn't there. The man was getting very annoyed because he wanted to vote and didn't know what to do now. I asked him for his voter ID and entered his number on the voter look-up tool on my phone. In a few seconds I got his entire information, and saw that he was at the wrong place—according to his record, he was supposed to be voting at a different booth. I told him where to go and he was quite grateful. Soon, the party workers were directing more and more people to me.

Party workers seized upon the look-up portal as a brilliant tool for making their job easier. All it required was one person with a smartphone; it was so easy to use that people were training each other on the fly, and eventually everyone was using the app by the end of polling day, irrespective of which political party they were affiliated with. Even by conservative estimates, we were able to help hundreds of people cast their vote that day instead of them being forced to return home without being able to exercise one of their most fundamental democratic rights.

Despite these and many other technological innovations that the campaign came up with, Nandan ended up losing the election to Ananth Kumar, the BJP candidate and six-time member of Parliament from Bangalore South. While Nandan polled more votes than the Congress candidate in the previous general election,[9] the campaign could not overcome the 'Modi wave' that swept the nation. The strong anti-incumbency sentiment that prevailed during the elections led many voters to tell us that Nandan was the right man, but in the wrong party. Ultimately, technology can only take you so far; it is the pulse of the people that finally decides which leader gets elected to office.

The party as a platform

Where do we see Indian elections evolving in the future? Truly speaking, political parties have always been brands, with a central message, a symbol, and loyal adherents. The party symbol and its ideology are inextricably linked, and based on these, people build expectations of what the party can do for them if elected to power. We believe that in the future, parties will begin to recognize and tap into the value of this brand, a process that has already begun. Rather than just offering a ticket and a symbol to the candidate, parties will now offer them an entire campaign outline, structured around the party's ideology and core strengths—what we refer to as the 'party as a platform' concept.

What will this platform consist of? For one, candidates will be given a strong central message which all of them can use, with some local customization. With their large budget, parties can ensure that the message stays consistent across the country, an approach that was most successfully employed by Narendra Modi and to a certain extent by the AAP in Delhi. Funds can be raised on the basis of the party's brand, and technology helps to log and utilize even the smallest of donations, as the AAP's fundraising campaign demonstrated so ably.

The platform will also offer a set of technology-based tools that candidates can access: digitized voter lists, voter look-up services for party workers, social media engagement, call centre management, 3D hologram rallies, and so on.

All these technological innovations were deployed in the 2014 elections, and should soon enter the mainstream of Indian politics. The trend of political parties increasingly providing a set of tools, or an integrated launch platform to their candidates, will only accelerate in future elections. Candidates that can combine strong grassroots support with a party-supplied platform will naturally have an edge over old-style politics.

Most importantly, we predict that future general elections will be 'mobile first'; over 500 million Indians will have smartphones in the next few years, and election campaigns will depend heavily on the mobile phone network to reach out to voters, even more than they do now. Parties that successfully implement the 'party as a platform' concept will have much more to offer to their candidates and voters, and will change the relationship between the party and the candidate. Today, parties offer candidates mindshare in the form of messaging and feet on the street. Once they can offer intellectual property in the form of software and a trained party base that can work with that software, the bargaining power between the party and the candidate will change in favour of the party. Such a strategy has been employed in other countries as well. In the US, every candidate of the Democratic party uses products from the technology company NGP VAN to manage their campaigns, whether it's an election to choose the next president or the head of the local labour union.[10]

In the Darwinian world of Indian politics, all the players will have to embrace and exploit the power and reach of technology if they want to survive.

11

THE ROLE OF GOVERNMENT IN AN INNOVATION ECONOMY

> The important thing for Government is not to do things which individuals are doing already, and to do them a little better or a little worse; but to do those things which at present are not done at all.
>
> —John Maynard Keynes, 1926

VETTATH SHANKARAN REPRESENTS a new breed of fisherman. Once upon a time, Shankaran used to rely on luck and experience to decide where to cast his fishing net on the high seas. Now, a network of satellites 20,000 kilometres above the earth feeds data into the Global Positioning System (GPS) installed on his boat, telling him where a good catch is likely to be found. In an earlier age, Shankaran would not be able to gauge where he could get the best price for his fish—if his fellow fishermen had also had a successful expedition, prices at his home market would fall. He could try sailing to other locations on the coast, but he had no guarantee that his fish would fetch him a better price there. Now, Shankaran and his associates simply call markets from Kottayam to Thrissur, negotiating the best price for their catch so that, 'by the time we reach the shores, our business deal has been settled', he says.[1]

The changes wrought by mobile telephones on the business

fortunes of Shankaran and his ilk were the subject of a landmark study by Harvard economist Robert Jensen.[2] He found that on average, consumers now paid 4 per cent less for their fish while the fishermen's average profit went up by 8 per cent; waste had been eliminated, markets had become efficient and fish were now uniformly priced across the region—what economists call the 'law of one price'. Follow-up studies found that fishermen credited their phones with more than just improving their income; they 'used their phones to maintain relations within and outside the market, and protected themselves during times of risk, vulnerability and emergency'.[3]

Certainly, when the Indian government threw open the telecom sector to private players to usher in mobile telephony, nobody could have anticipated that Kerala's fish economy would receive a substantial boost. But that the introduction of a new technology can have unexpected and beneficial consequences is a hallmark of the innovation economy. By throwing technology open to the public and bringing in players across the public–private spectrum, governments can spur growth and development in myriad new ways. Once, all telephony services in India were monopolized by state-owned corporations such as Bharat Sanchar Nigam Limited. Today, the government's primary role is no longer of a service provider with a cradle-to-the-grave approach, building telecom switches and handing out telephone instruments to subscribers. Instead, the government has defined rules for telecom companies, and established regulators like the Telecom Regulatory Authority of India, which ensure that laws are followed and consumer interests are protected. By taking a minimalist approach, in which private-sector participation competes in a regulated environment, the government has helped to make India's mobile phone network among the world's largest and cheapest.

Our marketplaces are in the midst of a revolution. The falling prices of smartphones and data services have led to new business models, which use the internet and information services to challenge existing players in fields ranging from retail to transportation. Traditional government domains are now being invaded by the private sector. The rise of the sharing economy, in which people directly share resources with one another, is further changing our understanding of business. In a time of such rapid change, government faces the challenge of implementing regulatory standards and protecting consumer rights in an ever-changing landscape. The mobile telephony network was among the first wave of technologies that required such efforts from the government. Let's see how some of the newer challenges are shaping up.

The rise of India's sharing economy

When Uber made its Indian debut in 2013, it became one of a clutch of taxi services debuting an unfamiliar business model. Instead of calling a central number, you could book a taxi using an app on a smartphone, track its location and arrival time via GPS, and pay via a credit card online. People were enthralled by the experience, and Uber experienced a surge in popularity.

However, this auspicious beginning was soon marred by regulatory troubles. First, the Reserve Bank of India objected to Uber's payment model, which violated the RBI mandate of a two-factor authentication for all credit card payments—designed to increase transaction security and reduce fraud. Uber initially managed to avoid this requirement by routing payments through a foreign gateway, since foreign exchange payments are exempt from the RBI's authentication rules. However, the RBI demanded that Uber either follow the same rules that applied to India-based taxi service providers or shut shop.[4]

Worse was to follow. In December 2014, an Uber driver in New Delhi was accused of raping a female passenger. The government responded by banning Uber in the capital region, and the events that followed conclusively proved that a 'service aggregator' like Uber, which claimed to only be a platform connecting drivers with riders rather than a commercial taxi service, occupied a regulatory grey zone.[5] Whose responsibility was it to ensure that Uber drivers had passed police verification checks? While the rules were clearer for commercial taxi services, what were the obligations for an aggregator like Uber, and could it truly claim to offer a safe and secure service to its customers? More troubling facts soon came to light. The driver was a habitual offender, but had managed to obtain a police verification certificate that authorities claimed was forged.[6] He had registered with Uber under a false name, and prior to assaulting his passenger, he simply switched off the app on his phone tracking the ride. Given that Uber was operating in India as a technology company rather than as a transportation business, whose job was it to monitor these safety norms and take action when they were flouted? Whether or not banning the service completely was warranted, it served no purpose unless it was also accompanied by the introduction of rules and regulations designed specifically to apply to service aggregators.[7]

Uber is not India's only service aggregator. Thanks to Airbnb, anyone can convert a spare room into a hotel without complying with hospitality industry regulations. eBay allows people to freely buy and sell goods from each other. All these services work as platforms, with buyers and sellers rating each other—a reputation that builds up over multiple transactions, making customers more inclined to use a service. Normally, this would require a regulator. Existing operators who bear regulatory costs—New York city taxi drivers who pay large sums to get a medallion, or hotels that meet regulatory requirements, making them more expensive—find disruptions such as Airbnb and Uber hard to compete with as they grow popular. Often, the new business models are different enough that they end up avoiding existing taxes. The incumbents label the newcomers as 'tax evaders' and the new solutions as 'unsafe'. There is also the risk of the existing interests lobbying against innovations and using regulations to create a barrier to entry. In all these cases, the conflict between regulation and innovation has led to considerable friction in the market.

The sharing economy is also giving rise to a new class of 'micro-entrepreneurs'—small-scale service providers who can now find customers through an online marketplace.[8] The US has seen the emergence of a plethora of start-ups that will bring you groceries and flowers, wash your clothes, deliver meals, clean your house, carry out your personal tasks, or send a doctor to your home if you're ill. You can hire a talented designer over the internet to create a logo for your company, or find consultants to tackle a challenging problem at work.[9] In urban India, you can now use apps to find cooks, cleaners and plumbers in your locality. There is a start-up for every possible opportunity that exists of organizing the unorganized. A recent report from the McKinsey Global Institute points out that 'India is a nation of small-scale, independent service providers' and by building digital marketplaces, these service providers can broaden their customer base and scale up for growth, whether they are graphic designers, nurses or carpenters.[10] Such collaborative platforms redefine the relationship between employer and employee, and the regulatory framework that supports this relationship—tax laws, employment laws, worker rights—needs to reshape itself.

The sharing economy and the services built around it are here to stay. One estimate places its current global value at $15 billion, a number projected to rise steadily.[11] It draws from both public and private resources, such as the internet, GPS and mapping services, as well as privately developed payment systems—credit cards, mobile wallets and the like, and depends heavily on establishing a culture of trust and reciprocity, whether that is through obtaining favourable reviews or conducting background checks when needed. We need to be ahead of the curve when it comes to building the regulatory frameworks that will allow such collaborative services to flourish while safeguarding the rights of the people.

Those who pioneer new business models in India will need to adapt our regulatory frameworks, given that the state's capacity to manage law and order and guarantee citizen safety may be inadequate. For example, taxi service providers may be required to build stronger safety platforms. This might mean that they run their own verification checks in addition to following the mandated requirements, ensure that only registered drivers operate vehicles, and that the tracking app on the driver's phone cannot be switched off while a ride is in progress. Some services have launched alert notifications, like Ola's SOS feature; in an ideal world, such a feature would be directly linked to the local police system. While transport service aggregators may be making the headlines right now for their regulatory troubles, it is naive to assume that such issues will not arise in other areas of business. The government needs to be prepared.

The public–private divide

In her book *The Entrepreneurial State*, the economist Mariana Mazzucato offers an interesting perspective on the state funding of innovation.[12] Agencies such as the US's Defense Advanced Research Projects Agency (DARPA), the National Science Foundation, the National Institutes of Health, the Department of Energy and others have a history of funding basic research at universities and research labs, often at a nascent stage where the risks are high enough to discourage investment from venture capitalists. This approach has helped to support some of the biggest technological and scientific breakthroughs of our time—consider that the development of the internet, the human genome project, nuclear technology, antibiotics and GPS have all benefited from early government support. Discussing the enormously popular products that came out of Apple under the leadership of Steve Jobs, Mazzucato says, 'The genius ... of Steve Jobs led to massive profits and success, largely because Apple was able to ride the wave of massive state investments in the revolutionary technologies that underpin the iPhone and iPad: the internet, GPS, touchscreen displays and communication technologies. Without these publicly funded technologies, there would have been no wave to ... surf.'

One particular example of the government funding of cutting-edge research is familiar to us. Along with his colleagues Jeff Bezanson, Alan Edelman and Stefan Karpinski, Viral is a co-inventor of the Julia programming language. Julia is an open-source, high-performance programming language under development since 2009. It is commonly used by scientists in the physical and social sciences, engineers and data scientists for diverse purposes ranging from exploring the secrets of the universe to teasing out new insights from big data. While Julia is itself an open-source project that received contributions from scores of programmers worldwide, it has benefited from government research funding at the Massachusetts Institute of Technology, from

US government agencies such as the Defense Advanced Research Projects Agency, the National Science Foundation and the Department of Energy. Given the fact that the commercial prospects of a new programming language at the point of conception are almost nil in today's world unless it is backed by a large company such as Google, Apple or Microsoft, it was unlikely that any investor would choose to invest in such a project. However, open-source contributions and government funding kick-started the project and have led to its maturity, to a point where universities around the world now teach Julia in classrooms, and companies use it extensively.

While government funding has been a key component of technological innovation, some of the resulting discoveries have ended up posing an unexpected challenge to government itself. For example, the era of the internet has helped to make public the kind of information traditionally restricted to the government domain, and the blurring of the boundary between governments and their citizens have led to conflict on both sides of the divide—the brouhaha around WikiLeaks is an excellent example, even though not all the information leaked was particularly sensitive or revelatory. When applications start to use this information for general use, new regulatory issues arise.

A particularly interesting case is that of Google Maps, a breakthrough when it was first introduced. Over the years it has become increasingly sophisticated, and hundreds of millions of people rely on it daily. But at every step, it has run afoul of various rules and regulations. The first contentious issue that Google Maps had to deal with was the highly charged subject of political boundaries. How does a mapping service accessible to anyone in the world display disputed international boundaries? Today, based on the internet address from where the request originates, Google displays a version of the map consistent with local law. Over the years, a number of issues have surfaced: the marking of sensitive areas on maps, such as military installations, and privacy and security concerns with the Google Street View service. Here we have a piece of technology that consumers love, but is simply too new and too innovative for regulatory processes to maintain pace. A knee-jerk response by many governments is to bring up sovereignty concerns, and ban innovation. For instance, the Survey of India has long enjoyed a government-endowed monopoly on the business of making maps in India, and we have seen them being openly hostile to innovative products like Google Maps. Private weather forecasters like Skymet and the Bangalore-based Citizen Weather Network are entering a domain that was until recently the exclusive preserve of the Indian Meteorological Department (IMD).[13] The holy grail of weather forecasting in India is predicting the onset of the annual monsoon season, and Skymet has already clashed with the IMD by releasing monsoon predictions and analyses that differ from the IMD's interpretation.[14]

Much like GPS, a more recent instance of a military technology being opened to the public is that of autonomous vehicles. Through DARPA—coincidentally the agency that also birthed the internet—the US government has been funding such endeavours for over a decade.[15] The underlying technology has now entered the commercial space; Google is testing self-driving cars using its Google Chauffeur platform, Uber has just announced an academic collaboration with Carnegie Mellon University to 'develop driverless car and mapping technology', and Apple is reportedly investigating technologies for building electric and self-driving cars.[16] While we may not see a fleet of self-driving cars taking over our streets in the near future, it's worthwhile to consider that various US state governments are already starting to pass laws that permit driverless cars to operate on state roads.[17] Once again, government regulations need to anticipate innovation by keeping a close eye on emerging trends and assessing their potential impact and chances of widespread adoption.

In another example, while space exploration has traditionally been a government purview, in recent years that has changed to include increasing participation from private sector companies and hobbyists. Thanks to the smartphone explosion, the prices of key electronic components and sensors—GPS, cameras, accelerometers and the like—have plummeted, making these easily accessible to the general public. The author and former editor-in-chief of *Wired* magazine Chris Anderson has referred to this phenomenon as the 'peace dividend of the smartphone wars'.[18] The accompanying diagram explains the payoffs from this dividend.

Today, a hobbyist could build a drone, mount a camera on it and take pictures of sensitive government facilities. A terrorist may use it for more dangerous purposes. More benignly, companies like Amazon want to deliver packages to their customers using small commercial drones.[19]

In the US, private companies like SpaceX are now pushing the barriers for spacecraft development and innovation, stepping firmly onto what used to be NASA's playing field.[20] As nanosatellites become cheaper, it's likely that more private operators will enter this space in India as well. A for-profit organization in India is now participating in the Google Lunar XPRIZE challenge to build and design a robot capable of landing on the moon, a domain that formerly only our government space agency, the Indian Space Research Organization (ISRO) operated in.[21] The entry of private companies into fields that have traditionally been the monopoly of government will no doubt create a new set of regulatory clashes to be ironed out.

Innovation within government: Aadhaar as an hourglass

Maintaining a balance between innovation and regulation was something that we dealt with on a daily basis during our days at the UIDAI. The energy futurist Amory Lovins, chairman of the Rocky Mountain Institute and a personal friend of Nandan's, gave a talk in Bengaluru in which he discussed the four types of innovation that are essential to bring about any large-scale change in society: policy, technology, design and strategy innovation.[22] The Aadhaar project checked all these boxes. Policy innovation was required to allow Aadhaar and its allied services, such as e-KYC, to work. The development of biometric technology at Indian scales and price points, including the ability to carry out de-duplication for a billion people, the design and rollout of the enrolment programme and other Aadhaar-based applications—all these aspects of Aadhaar met Lovins's definition of innovation.

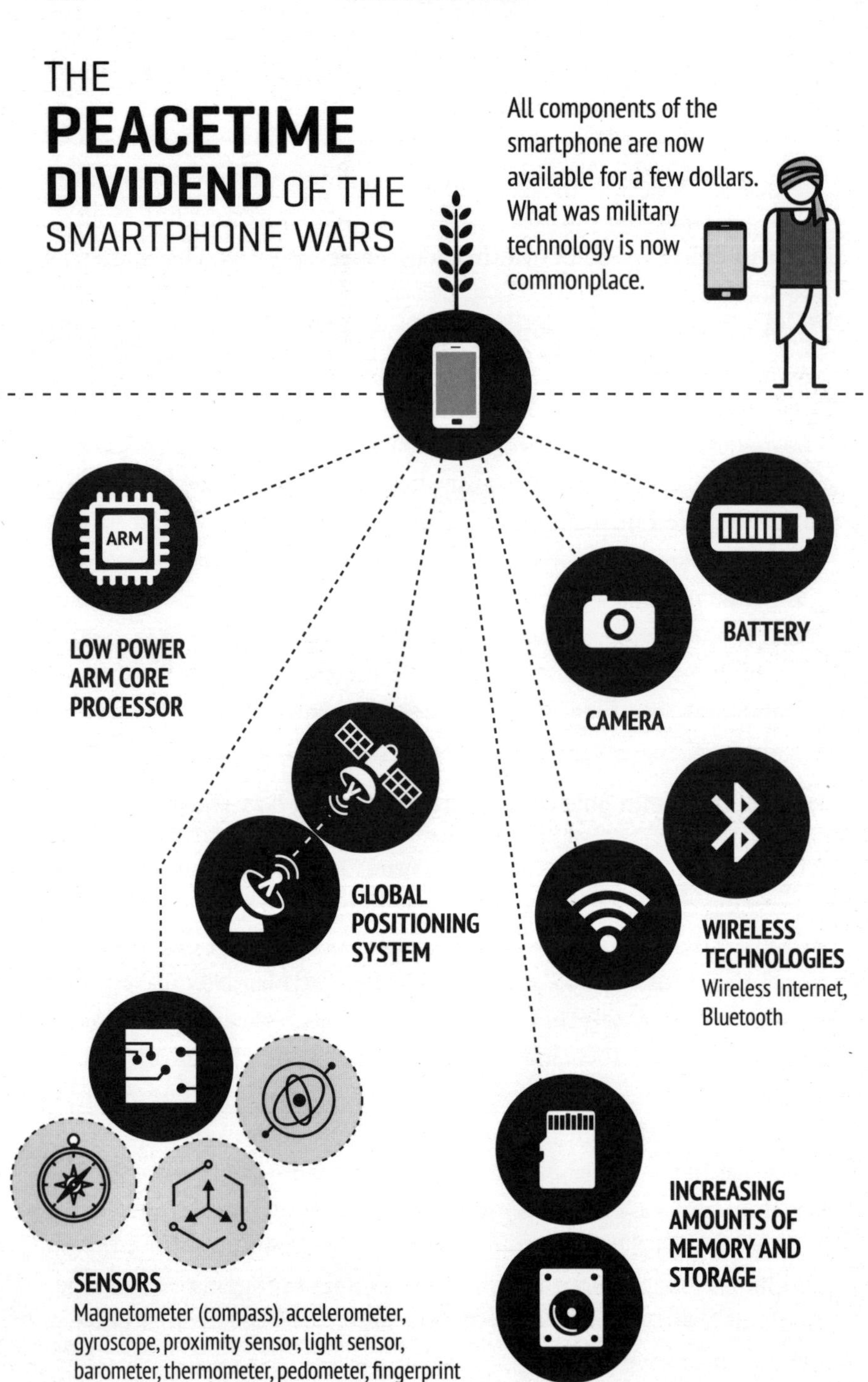
THE
PEACETIME
DIVIDEND OF THE
SMARTPHONE WARS
All components of the smartphone are now available for a few dollars. What was military technology is now commonplace.
ARM
LOW POWER
ARM CORE
PROCESSOR
BATTERY
CAMERA
GLOBAL
POSITIONING
SYSTEM
WIRELESS
TECHNOLOGIES
Wireless Internet,
Bluetooth
INCREASING
AMOUNTS OF
MEMORY AND
STORAGE
SENSORS
Magnetometer (compass), accelerometer,
gyroscope, proximity sensor, light sensor,
barometer, thermometer, pedometer, fingerprint

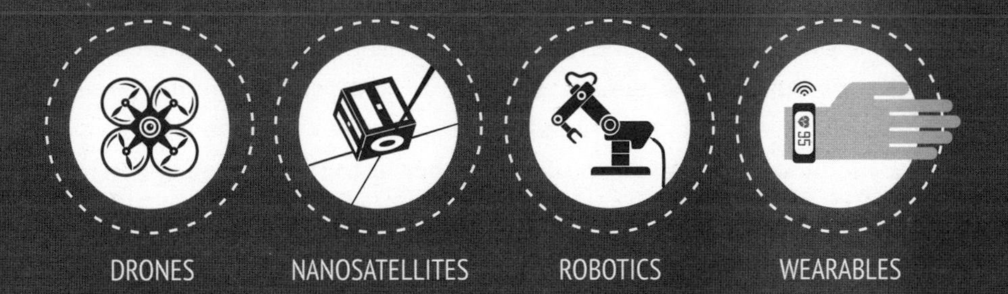

These have led to multiple new personal products and entire new industries

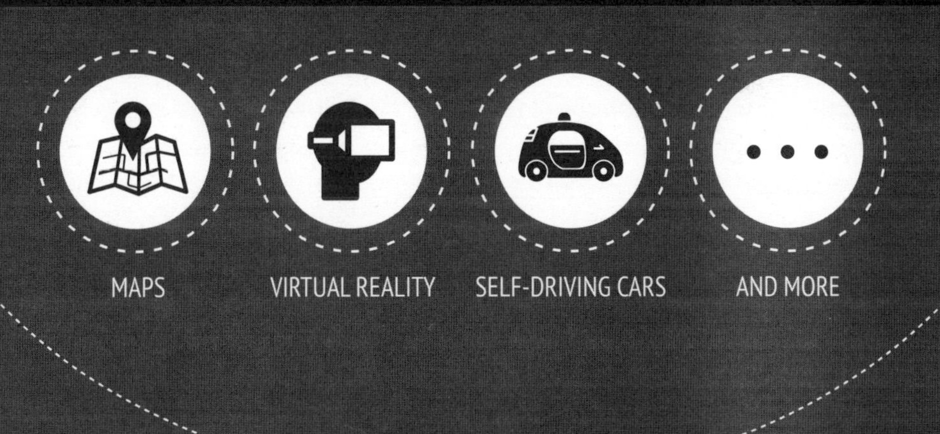

Each of these innovations clashes with current laws and regulation

Airspace is tightly controlled, on what can fly and who can fly it

In India, Survey of India has a monopoly on maps

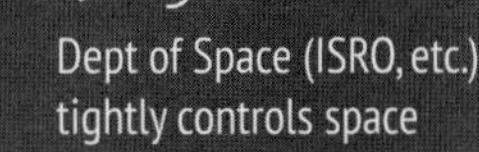

Dept of Space (ISRO, etc.) tightly controls space

THE CHALLENGE:

Technological innovations improve people's lives but clash with the government's notions of security and maintaining law and order. How will the government encourage innovation while providing regulation?

It wasn't until much later that we realized that Aadhaar also fit neatly into another model of innovation, one that originated as a technology concept but is now gaining widespread acceptance. In late 2014, Nandan gave a talk at the Massachusetts Institute of Technology on the architecture of Aadhaar. One member of the audience happened to be Dr Chintan Vaishnav, whose research focuses on the intersection of technology and government policy. Upon hearing Nandan's talk, he pointed out that Aadhaar's design was strikingly similar to what has been christened the 'hourglass model', first mentioned in a 1994 report by the National Research Council in the US. As Chintan explains, 'The "hourglass architecture"—where minimal standardization at the waist of a layered architecture (e.g. the IP layer of the internet) enables burgeoning innovation above (e.g. apps) and below (e.g. wireless, wired networks, etc.) it—has been key to the innovativeness of the internet as a platform.'[23] In other words, a simple, easy-to-use solution forms the waist of the hourglass, while allowing for innovation in multiple spheres both above and below. One such example is the GPS technology that helped to change Vettath Shankaran's fortunes. While the GPS technology itself can be considered the narrow waist, below it lie any number of devices and sensors that can detect this signal, and above are a suite of applications—taxi services, fitness apps, and so on—that use GPS services to determine location for a multitude of uses.

How does this model apply to Aadhaar? Here, the waist consists of the Aadhaar number—a unique identifier for every individual—and authentication services linked to this number. Below the waist lies innovation in design, in this case biometric devices that can capture fingerprints and iris data. It also includes innovations in the way people are enrolled for Aadhaar. Above the waist lies any application that might require an identity verification service. That number has already hit half a dozen and is steadily inching upwards; applications that require ID verification currently include the central government's biometric attendance system, Aadhaar-based payments, direct benefit transfer for subsidies, social security payments, PDS reforms, KYC for bank accounts and SIM cards, and more.

It turns out that the hourglass model is a good approach to building governance systems. It forces us to think about the minimal intervention that government needs to make in order to provide public goods and services. What may appear to be a simple solution is the outcome of assembling several experts, understanding the domain and precisely defining the 'thin solution' that will make the delivery of government services more efficient while also encouraging the development of innovative solutions to long-standing problems.

As the accompanying diagram explains, focusing on the waist of the hourglass—the thin solution—gives government a significant competitive edge. Rather than having to build entire systems from the ground up, government can choose to concentrate on creating a single, specialized system that delivers one specific outcome, an approach that also lends itself well to being scaled up. This single solution can be widely compatible with both public and private services, allowing an entire ecosystem to spring up with little additional effort on the government's part, and creating a structure that rewards innovation. From the political standpoint as well, a thin solution makes sense—by choosing to offer only a single, tightly defined service, government departments can avoid stepping on too many toes within the administration as well as reduce the amount of opposition from external sources. Thin solutions don't require drastic changes to the status quo, further reducing the potential to ignite the kind of turf wars that have already sunk plenty of well-intentioned government initiatives. They are also easier to write into law since they are restricted in scope, and since they are easy to implement and scale up, they can gather enough momentum that they become irreversible.

A new class of government institutions

If we are to build a new class of institutions capable of nimble, creative thought and action to promote innovation in government, we need regulatory innovation to occur in lockstep. The concept of a regulator is derived from the field of electrical engineering; a voltage regulator tries to maintain electrical voltage at a constant level, protecting the

DESIGN THE IDEAL GOVERNMENT SYSTEM

1. **NARROW & SPECIALIZED**
   Do one simple thing and do it well.
2. **MINIMAL INTERVENTION**
   Don't make too many changes.
3. **EASY TO WRITE IN LAW**
   Simple laws are passed easily.
4. **MINIMAL OPPOSITION**
   Factor in diverse viewpoints.
5. **EASY TO SCALE**
   Execution is easy to understand, requires little training and can be repeated millions of times.

THE THIN APPROACH — **Simple design** that allows innovation on all sides

EASY TO EXECUTE

EASY TO WRITE THE LAW

EASY TO BUILD THE TECHNOLOGY

6. **IRREVERSIBLE**
   Operate at large scales to make projects permanent.
7. **MINIMAL TURF ENCROACHMENT**
   Fewer toes to step on.
8. **BUILD AN ECOSYSTEM**
   A level playing field for both governments and markets.
9. **STRUCTURE FOR INNOVATION**
   Provide simple solutions and let markets drive innovation.

equipment connected to it from sudden fluctuations. In the same fashion, regulatory bodies are meant to buffer the population they serve from the vagaries of the open market. In modern times, regulators are also expected to devise policies that encourage innovation. Reconciling these two opposing goals requires a delicate balancing act. For example, unchecked innovation in financial products led to the financial crisis of 2007. While regulation may sometimes lag innovation, it should not curb it.

After the economic liberalization of the 1990s, we have built a number of regulatory institutions. Regulators that oversee the financial sector include the RBI, the SEBI and the IRDA. Legislation was also passed that enabled the creation of electronic depositories to hold securities. The NSDL paved the way for dematerialization of share certificates, an essential component of electronic trading; today, NSDL is one of many flourishing stock depositories.

These institutions are likely to be restructured and strengthened under the recently established Financial Sector Legal Reforms Commission. We also have regulators for pharmaceuticals, electricity, food, and for monitoring the anti-competitive behaviour of firms in the market. In order to function optimally, these regulatory bodies need individuals with expertise in relevant areas, but haven't been able to attract the kind of talent they need. Government officers who have spent decades in administration are simply not on the cutting edge of technology and are often unaware of the rapidly shifting landscape around them.

We desperately need to rethink the human resources policies for our regulatory bodies so that they can deliver on the twin goals of safety and growth. It takes a mature government to create an ecosystem conducive to growth and change, and to support it with the appropriate regulation. Aadhaar and its linked suite of applications would never have come to fruition if various regulators had not understood the potential of empowering every resident of the country by granting them a unique ID. With a bold vision and some regulatory muscle to back it up, we remain confident that Aadhaar will be the first of many schemes that completely reshape the relationship between the Government of India and the people it is meant to serve.

A new approach to bureaucracy

A senior ex-bureaucrat explains to us the unintended consequences of the checks and balances imposed by the government upon civil servants:

> I was in a senior post at the State Farms Corporation of India some years ago. We were told about some unexpected rainfall which was likely to happen soon, and we decided to buy gunny bags to protect the stored seeds from getting spoilt. We didn't have time to follow the standard procedure of publishing a tender in the newspapers and identifying the lowest bidder. The marketing officer suggested that we just buy them from the same supplier we usually used. Much after the rain had come and gone, a question was asked as to why we had not published a tender; was there some sort of underhand agreement between the marketing officer and the supplier? Of course, we were able to justify our actions, and the matter was closed, but had a serious inquiry been launched, the matter would have left a black mark on the officer's career for no fault of his.

The fear of being dragged in front of an inquiry committee tends to make bureaucrats risk-averse and in doing so, stifles innovation.[24]

While it's all too easy to deride our bureaucracy as being out of step with society, it is important to remember that the incentive structure under which the bureaucracy operates is completely different from what we may encounter in the private sector. In his book, *Bureaucracy: What Government Agencies Do and Why They Do It*, James Q. Wilson explains in wry detail the competing goals that government agencies must meet: they are expected to fulfil an official mandate laid out for them by the government as efficiently as possible while also treating all citizens fairly, maintaining a certain level of responsiveness to citizen concerns and acting in a fiscally responsible manner at all times since they are accountable to both the government and the public for every rupee spent. These goals can, and do, often clash. Add to this the amount of time that officials must spend securing government support for various initiatives, and it is not surprising that many government projects are executed in a less than ideal fashion. Building a new class of government institutions will remain a pointless exercise unless we also create new ways of staffing these institutions with the best available talent, a point that was addressed in the 2013 report of the Expert Committee on the HR Policy for e-Governance, a committee chaired by Nandan.[25]

Throughout this book we have discussed the need for a new operational model of government, the National Information Utility, as being essential to implement technology-enabled solutions to the challenges our country faces today. Such a model must necessarily draw on talent from both sides of the public–private divide, and we must understand and appreciate the strengths that both sides bring to the table. Government employees have a deep understanding of procedures, protocols and how government systems work. They are focused on stability, and on executing projects which can stand the test of time after being implemented on a vast scale. They also understand that government projects are designed to be inclusive, providing benefits to as large a section of the population as possible, rather than being designed to provide a very high level of benefits only to a select few.

On the other hand, the private sector has traditionally focused on profit-making as a goal, which has meant the development of a larger appetite for risk and greater comfort with uncertainty and shifting goals. A high value is placed on the ability to understand and employ technology not just to answer today's problems, but tomorrow's as well. Academics and those from non-governmental organizations also have their own, equally valuable perspectives. All of these different viewpoints and experiences must be harnessed and integrated to provide holistic solutions, explained in further detail in the diagram that follows. Some of these boundaries are already starting to blur.

WHAT **GOVERNMENT** IS GOOD AT

SCALE
Operates at nationwide scale

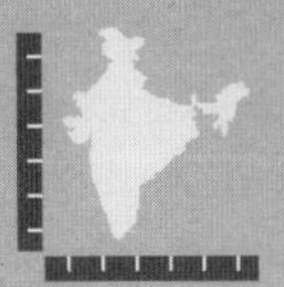

STABILITY
Runs reliably for decades

STAFFING
Administrators who handle every type of project

INCENTIVES
Based on process rather than outcome

WHAT **A START-UP** IS GOOD AT

SPEED
From design to market in months

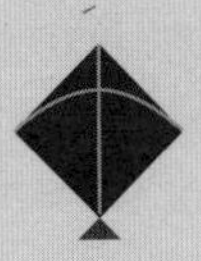

AGILITY
Build fast, fail fast until the right solution is found

STAFFING
Domain experts who solve a specific problem

INCENTIVES
Shaped by the market

WHAT **WE NEED**

1. Combine the scale of government with the speed of start-ups

2. Merge stability and agility

3. Bring together administration and domain expertise

4. Leverage the power of markets while adhering to due process

Raghuram Rajan, the current governor of the RBI, is a former academic; his appointment was widely hailed and even credited for halting the precipitous slide of the rupee against the dollar.

From friction to harmony: What Aadhaar can teach us

Drawing as it did upon expertise from both the public and private sector, the UIDAI could be considered an early forerunner of the NIU, and our experiences provide some idea of the ups and downs such an organization could expect to face. We saw first-hand the effect that an ambitious goal can have on bringing people together to work at their best. As Pramod Varma, Aadhaar's chief architect, puts it, 'There was respect for each other, there was a larger goal which brought in the purpose. Issues were overlooked because we were there for a larger goal. Because it was a start-up-like environment, even for government officials, the sense of purpose connected people.' Speaking from the perspective of his government background, Rajesh Bansal adds, 'The only way to get such projects to succeed in government is getting people to take ownership, and passion is the incentive for that. Government recruitment processes don't usually focus on passion for the job as a key criterion.'

Inevitably, some friction arose, and resolving these tensions turned out to be an invaluable learning experience in terms of getting the organization to run smoothly. Much of this friction centred around who ultimately had the power to take official decisions within the UIDAI. Ram Sewak Sharma tells us, 'Government officers felt that this is a government project and they should have full control on policies and decision-making. On the other hand, resources from the private sector felt that they had been hired to deliver and they should also have the ownership.' Srikanth Nadhamuni explains, 'The technology team was holding the vendors on a tight leash, but the fact that the payments were coming from the government gave out mixed signals—who was the vendor really working for? It would have been much easier if a government official had been deputed to work in tandem with us for handling the payments angle—we could have worked together to resolve technical issues as well as the government payments process.' Of course, this did happen eventually.

Government comes with its own set of rules and standard operating procedures, many of which are completely unfamiliar to those outside the administrative orbit. Shankar Maruwada says, 'We had to understand that bureaucrats are primed to follow process and not take risks, which was a big change for us. They have a massive number of rules to follow and we learnt to appreciate the constraints under which they work. We also had to learn that in government, you have to wait for the correct window of opportunity instead of rushing things—the fact that you're not moving doesn't mean there's a lack of progress. Understanding this taught us a lot of patience.'

A common theme was the need to create official positions for those brought laterally into government agencies from the private sector, positions with well-defined powers and responsibilities. As Ram Sewak Sharma puts it, 'The lesson for me is that we should clearly articulate the roles of people, including the hierarchy, for any government organization in India to work efficiently.' Ashok Pal Singh adds, 'You need to provide the flexibility and freedom to function while remaining within the purview of the government.' This freedom was in evidence at the UIDAI. Pramod recollects, 'As technologists, we had the kind of large-scale, hands-on product experience that does not commonly exist in government. Members of the initial founding team like R.S. Sharma recognized this and gave us a lot of leeway and support when it came to working out the details.'

Ram Sewak Sharma summarizes it for us, 'Perhaps we could not achieve perfect harmony between the public and private sector at the UIDAI—five years is a short tenure to change the kind of mental conditioning created by the working environment, whether in the government or outside it. But we need to also remember that this project could not be delivered either by private or by public team alone, and what we did was the only way we could have done it.'

12

TOWARDS A HEALTHY INDIA

> When health is absent, wisdom cannot reveal itself, art cannot manifest, strength cannot fight, wealth becomes useless, and intelligence cannot be applied.
>
> —Herophilus

IN THE SEVEN months of Sarman Siddiki's brief life, his mother, Moksuda, took him to various government hospitals across the state of West Bengal, trying to get her son's persistent diarrhoea cured. They lived in a small village two hours away from Kolkata, and Moksuda travelled in trains and buses to get medical care for her child. Many doctors sent Sarman away with prescriptions for medicines; he was admitted to one hospital, sharing a packed ward with children suffering from other contagious diseases. His family spent an entire month's salary on their son's treatment, but took him home when it was clear that the overburdened medical staff had little time for him. Sarman's journey ended at a crowded government hospital in Kolkata, where he became one among forty-one babies who died in a span of six days.[1] This is the same city in which the first recorded study of diabetes in India was performed in 1938. At that time, only a thousand of nearly one hundred thousand individuals were found to have the disease; thanks to the twin imperatives of globalization and urbanization, today over half a million residents of Kolkata are diabetic.[2]

Nowhere are the paradoxes of India's rapid economic rise encapsulated better than in the health of its residents. As a nation, we are caught between the devils of our underdeveloped past—malnutrition, high maternal and infant mortality, communicable diseases like tuberculosis and malaria—and the deep sea of sudden prosperity, with an alarming rise in obesity, heart disease, diabetes and other lifestyle-related illnesses. We currently have the world's second-largest diabetic population, while nearly half of our children continue to be chronically malnourished.[3]

India has suffered from years of public health policies so inert as to be practically comatose, developments that Nandan traced in *Imagining India*. The situation continues to be grim. As per the Economic Survey of 2013–14, India's central government spent a measly 1.4 per cent of GDP on healthcare, compared to 3 per cent in China and 8.3 per cent in the United States; our public health spend is among the lowest in the world, and only 33 per cent of this expenditure is funded by public sources.[4] The standard of care in government-funded public hospitals varies wildly across the country; the ratio of the number of doctors and hospital beds available to the total population is far below the benchmarks set by the World Health Organization. The lack of easily accessible, high-quality public healthcare has led the private sector to pick up the slack, with private hospitals and clinics mushrooming over the last decade. Faced with ramshackle, poorly staffed and underequipped government hospitals, it isn't surprising that people make a beeline for private medical facilities even if they are more expensive; over 60 per cent of Indians pay medical costs out of their own pockets.[5] For India's poor, a single health emergency can be sufficient to ensure a lifetime of crippling debt, derailing attempts to overcome poverty; according to the ministry of health, 63 million people every year are faced with 'catastrophic' expenditure due to healthcare costs.[6]

India's public healthcare system is in desperate need of a massive overhaul. While our existing medical facilities, especially in rural areas, are inadequate to serve the needs of the population, they are also underutilized by those who prefer private healthcare. Reforming our healthcare system will require changes in policy—including a greater investment from the government—upgrading existing facilities and expanding the reach of public hospitals, clinics and primary care centres.

Broadening reach and cutting costs

One way to reach underserved and remote rural areas is by providing medical services over the internet or by phone. India's private sector has already seen the implementation of telemedicine facilities across hospital chains, allowing doctors to directly interact with and diagnose patients sitting thousands of kilometres away.[7] Such facilities can serve as the first point of contact for patients with the healthcare system, and can help to address the problem of building and staffing healthcare clinics in hard-to-reach parts of the country. Another exciting development that must be harnessed is the ability of smartphones to function as diagnostic devices. A cursory search through the online app stores of mobile telephone systems reveals a profusion of apps designed to allow people to monitor their heart rates, track how many calories they've burned, and help them manage chronic conditions such as diabetes. A new wave of research is now focused on developing small, mobile phone-compatible sensors that can be used to measure physical parameters, such as a person's blood glucose level. By simply plugging the sensor into one's phone, the relevant data can be stored, analysed and transmitted as needed to healthcare professionals. An entire suite of such plug-and-play sensors can allow for multiple health parameters to be monitored simultaneously, and given the size of India's cellphone user base, could convert even the most basic of primary health care centres into rudimentary diagnostic laboratories.[8]

While broadening the reach and quality of medical facilities within the country, a second priority for our healthcare system must be focused on lowering costs. A report from the consulting firm Deloitte estimates that between the years 2005 and 2015, the economic impact of heart disease, stroke and diabetes on the Indian economy was to the tune of over $200 billion in losses—an amount equivalent to 1 per cent of our GDP, putting an enormous strain on our finances.[9] The best way to allow for efficient, low-cost patient management is through the creation of electronic health records (EHRs). Today, the average person's medical history is usually encapsulated in a bulging file full of doctor's notes, test results, X-rays and scan reports stretching back several years. Medical service providers usually have their own internal data storage systems, which operate completely independent of each other, so it is impossible to track the progression of a particular patient through the medical system.

The surgeon and author Atul Gawande describes an interesting experiment carried out by Jeffrey Brenner, a physician in Camden, New Jersey.[10] Brenner started monitoring the flow of patients into Camden's hospitals, and focused his attention on those patients who cost insurers the most money. As it turns out, these were not people having complex medical procedures or being given expensive medicines—they were people whose illnesses were not being managed correctly, like the man who wasn't taking his blood pressure pills and was headed for kidney failure as a result. Something as simple as checking to make sure he took his medicines on time was sufficient to improve his health and save the insurers from paying thousands of dollars in dialysis costs.

In the Indian context, Brenner's model has plenty of applications. For instance, the indiscriminate use of antibiotics and the failure to follow treatment plans has led to the emergence of drug-resistant forms of tuberculosis and other 'superbugs', a looming crisis that we are completely unprepared to handle, both medically and financially. If one could monitor all the antibiotics that a person has been prescribed over time, and make sure they have been taken correctly, the problem of antibiotic resistance might be averted. This kind of health monitoring is only possible if every patient has an EHR in a common, standardized and interoperable format. Pioneering initiatives already exist in India, such as the EHRS system implemented by the Apollo chain of hospitals.[11]

Digitizing health: An electronic medical record system

What are the benefits of creating a common, nationwide platform for healthcare? Patients can have easy access to their medical history, without having to save every scrap of paper from every doctor's visit. After putting in place rules governing privacy, data sharing and access, this data can now be made available to all healthcare providers who can make better-informed decisions. Machine-learning algorithms can be applied to come up with new insights and develop a customized treatment plan for every patient. This data can also be mined to understand larger trends in public health across the country.

In terms of its physical design, we envision that an Electronic Medical Record System (EMRS) will be a service provided by multiple technology service providers. Every provider will conform to a set of interoperability guidelines, allowing users of the platform to pick a vendor of their choice, as well as making it easy to move from one service provider to another. The EMRS will provide interfaces to all stakeholders: hospitals, pharmacies, clinics, insurance companies, diagnostic labs, and so on. For example, a pharmacist can pull up a prescription, doctors can pull up diagnostic test results from labs online, and insurance companies can provide customized quotes based on prior history. The Aadhaar number can serve as the natural patient identifier in this system, and a base level of information can be shared automatically, while the release of additional information will require patient consent—a 'part public, part private' model.

Given the sensitivity of the data it holds, an EMRS system must be highly secure and must meet all requirements of patient privacy and data confidentiality. Given the emergence of big data and cloud computing, all records can simply reside online in the cloud, allowing for easy storage and access. The design of the entire EMRS must respect the fact that the medical data of a patient belongs to them, and they can take it with them wherever they want, or download it for personal use. EMRS providers that focus on customer convenience through such services as smartphone access, offer a simple user experience, have a good track record, and offer competitive pricing, will end up winning the patients' trust and their wallets.

There are natural questions about what data is stored, and who is allowed to access it. A strong data regulator is essential. In the US, the Health Insurance Portability and Accountability Act (HIPAA), 1996, regulates data sharing and privacy; we need a similar data regulatory body to be set up by the government, whose duties would include the preparation of legislation to govern data sharing and privacy, the laying down of interoperability guidelines and file formats for data sharing based on global standards (such as the HL7 and DICOM standards that govern medical data), and the monitoring of EMRS providers for compliance on an ongoing basis.

The EMRS will be a treasure trove of 'big data' that can be mined using analytics to identify public health trends, collect statistical data, perform disease surveillance and detect epidemic outbreaks. Eventually, this information can be combined with other social and environmental data to drive the sort of holistic approach to preventive healthcare that helped to improve quality and cut costs in Camden.

India's government-run medical facilities are notorious for the absenteeism of staff; a World Bank survey reported that an estimated 40 per cent of primary health clinic staff were missing in action at any given time.[12] EMRS would help to monitor the attendance and performance of government healthcare workers in primary health centres and hospitals, as well as Accredited Social Health Activist (ASHA) and Anganwadi workers and other paramedical staff. The EMRS can also be used to track government-run insurance schemes (the Rashtriya Swasthya Bima Yojana scheme is a prime example) by collating data from the EMRS and empanelled insurance providers.

This is an idea that Nandan had mooted with several health secretaries during his time at the UIDAI, but it failed to gain any traction within the administration.

Health on the cloud: A National Health Information Network

Building an EMRS is a long-term investment, requiring the creation of new legislation, new regulatory bodies and perhaps a new National Information Utility (NIU). In our road map to a healthy India, we envision this NIU to take the form of a National Health Information Network (NHIN), a common platform to bring all players in the health ecosystem under one digital umbrella, paving the way for the creation of a paperless national health network. Given the proliferation of health-linked smartphone apps and wearable diagnostic devices, the NHIN should function as a resource to promote further innovation in these fields by acting as the central repository for all health data in the country.

As all government hospitals and primary healthcare centres come online, the government will be the first driver of the NHIN, while other players can enter the system over time—an asynchronous design that allows the NHIN to be launched rapidly without having to spend an interminable amount of time ensuring that every single participant is enrolled prior to the launch. In the same way that Aadhaar-linked bank accounts can be used to disburse government payments, the Aadhaar-linked NHIN can be used by the government for health-related payments to individuals, such as benefits through the Janani Suraksha Yojana or incentives for ASHA workers.

Eventually, we expect the payment horizon to encompass the private sector as well; for example, insurance companies that need to pay health service providers under the terms of an insurance policy can use the NHIN to do so. Skilled professionals who understand both medicine and technology will be key in the implementation and rollout of such a system. The ministry of health can create a new department to house a dedicated team with the mandate to build the NHIN ecosystem.

We as a nation can no longer afford to be blasé about short-sighted policies, lopsided development and inadequate funding when it comes to healthcare. With public facilities in a shambles and private facilities unaffordable to the poor, simultaneously afflicted with diseases of

1 Every resident will have an Aadhaar, a smartphone and a bank account.

ROAD TO A HEALTHY INDIA

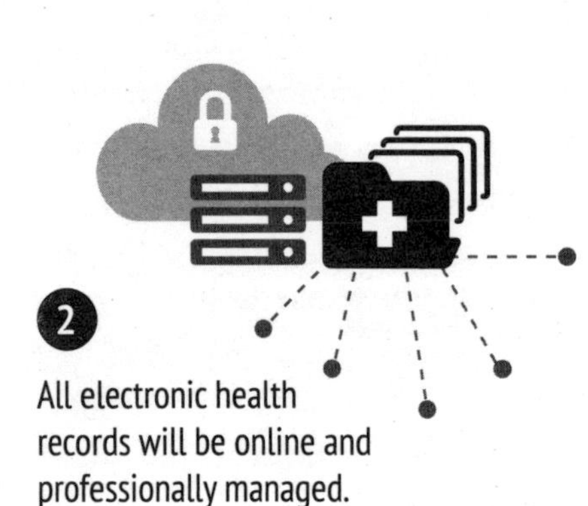

2 All electronic health records will be online and professionally managed.

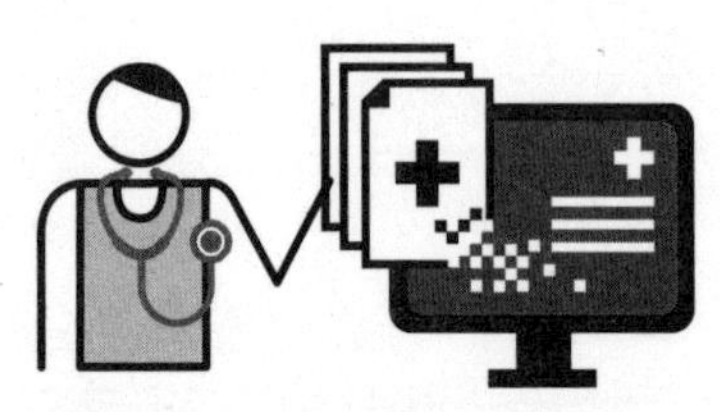

3 All players in the medical system will join an electronic health network.

4 A data regulator will manage privacy and security issues.

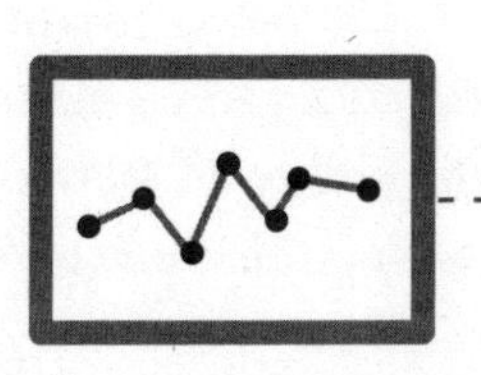

5 Data from an electronic network can be used for public health planning – monitoring disease outbreaks and measuring health metrics.

6 Smartphone apps and wearable devices will change the way health data is collected.

7 Data from smartphones and wearables will be integrated into electronic health networks.

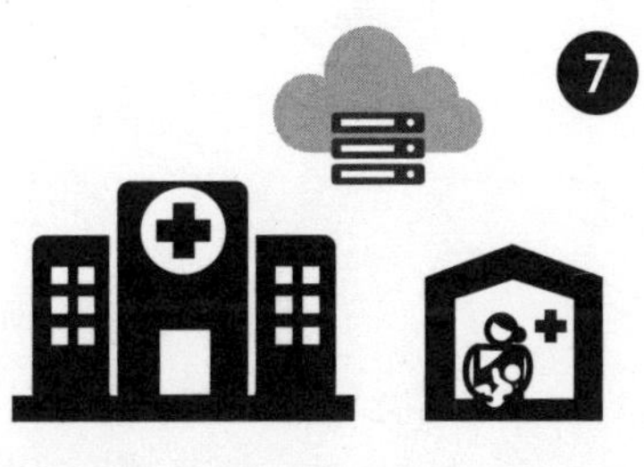

8 Government hospitals and primary health centres will be the first entrants into an electronic health network.

9 Government health welfare schemes and insurance companies will route payments electronically into Aadhaar-linked accounts.

poverty and diseases of affluence, our economy is being insidiously weakened by a workforce that is growing sicker. Piecemeal solutions will no longer work, and we must harness the ability of technology to give us the big-picture view essential to solving our most pressing health crises.

13

TEACHING THE NEXT GENERATION

The mind is not a vessel to be filled but a fire to be kindled.

—Plutarch

IN FEBRUARY 2015, the state of Maharashtra held its annual evaluation tests for nearly 400,000 teachers of government-run schools, who taught either primary school (classes 1–5) or upper primary school (classes 6–8). Of all the primary school teachers taking the test, only 1 per cent passed. Their colleagues in the upper primary section fared better—nearly 5 per cent of them passed the exam. To put it bluntly, 99 per cent of Maharashtra's primary school teachers and 95 per cent of its upper primary school teachers are unfit to teach, and this is in a state that has better education statistics than most—99 per cent of children between the ages of six and fourteen in Maharashtra are officially enrolled in school, and there is one teacher for every twenty-five students.[1] These alarming test results belie the fact that teachers in India receive up to 80 per cent of all public expenditure on education.[2]

'Few things are as wide-ranging in their impact on the economy as education. The collapse of our schools is a deep crack in India's foundation, and it impacts everything from our health achievements and fertility rates to our economic mobility and political choices.' This statement from *Imagining India* emphasizes the vital role of education in our nation's progress. Nearly 25 per cent of our population is still illiterate;[3] how can we hope to reap the benefits of the technological revolution if one in four people cannot read or write? Our counterparts in the group of BRICS nations all have better literacy rates—nearly 90 per cent—and better student-to-teacher ratios than we do.

Rising inputs, falling outcomes

The problems bedevilling India's education sector are not new, reflecting the impact of a complex set of social and political ideologies whose historical evolution Nandan has already traced in *Imagining India*. Since then the government has continued to launch new schemes designed at promoting universal education. First came the Sarva Shiksha Abhiyan, followed by the Right to Education Act (2009), both laudable pieces of legislation in their vision and scope for educating India's children.

Sadly, as is true of so many well-intentioned schemes in India, implementation has been the big stumbling block. Focused on easily measurable outcomes—schools built, teachers appointed, attendance rates of pupils—the quality of the education being provided has taken a back seat. As Rukmini Banerjee, the director of the Annual Status of Education Report (ASER) centre, says, 'By just providing inputs we are not bringing in equity. Unless the entire expenditure and the effort behind the provision of schooling is translated effectively into learning outcomes, the real battle for equal opportunity will be lost and our large and growing public expenditure in education wasted.'[4]

While enrolment rates remain high, many students drop out within a few years of entering school; UNICEF estimates that out of the nearly 200 million children enrolled in school in India, 80 million are likely to drop out without completing elementary school.[5] Even while they're in school, they're not learning much; according to the latest ASER published under the aegis of Pratham, an NGO that has led pioneering efforts in education reform, 78 per cent of children in standard three and 53 per cent of children in standard five cannot read a standard two-level text.[6] The study has also uncovered an escalating preference among parents to enrol their children in private schools, even in rural areas, believing that private schools offer a higher standard of instruction and are a better investment towards securing their children's future. These are the outcomes after India has spent $94 billion on primary education over the last decade.

The growing split between public and private schools is symptomatic of the larger malaise affecting the field of education. As Madhav Chavan, Pratham co-founder, states, 'The dominant thinking in the education establishment for the last decade has been that if we do more of what we have been doing and do it better, the quality of education will improve.'[7] Clearly, this mode of thinking has to be jettisoned, and fast, if we are to catch up with our goals of educating our children. Karthik Muralidharan, an economist and professor at the University of California, San Diego, writes, 'It is therefore imperative that education policy shift its emphasis from simply providing more school inputs in a "business as usual" way and focus on improving education outcomes.'[8] If we are to effect rapid and scalable change, we need to use the kinds of technological tools that are starting to revolutionize the way people disseminate and accumulate knowledge around the world.

Flipping the classroom

Online education systems are disrupting the traditional learning experience at both the school and college level. At the school level, teacher and entrepreneur Salman Khan's Khan Academy, a free online education platform, is changing the way students work in a classroom. Khan Academy provides online videos on a variety of subjects, along with problem sets whose difficulty level changes depending upon the capabilities of the user. By viewing these videos at home and then working on homework during school hours with the assistance of a teacher, Khan's videos are flipping the traditional classroom set-up.[9] In Khan's words, 'Students can hear lectures at home and spend their time at school doing "homework"—that is, working on problems.

It allows them to advance at their own pace, gaining real mastery, and it lets teachers spend more time giving one-to-one instruction.' Khan is now working with schools to integrate his videos into the standard curriculum. His team also created a dashboard that would allow teachers to monitor students' progress at an individual level, an idea that sounds simple but has tremendous potential. 'We'd go collect some data and make a chart, and the teachers were blown away—every time,' he says. 'This isn't taxing the edge of technology. But they were completely shocked, as if this had never existed before.' By creating a model in which students can learn at their own pace and are rewarded for mastery of a certain topic while also allowing teachers to monitor progress and provide targeted help to each student, Khan Academy has pioneered a practical approach to the concept of one-on-one education.

Technology can also be used to address what Karthik Muralidharan has termed 'the biggest crisis in the Indian education system'—the challenge of providing high-quality primary education.[10] In addition to the statistics we quote above, ASER surveys show that only a quarter of all students in standard five can perform simple mathematical division, and less than half can read. The magnitude of the crisis in primary education calls for out-of-the-box solutions that can help to fill in the gaps that the current system cannot address. To this end, Nandan and his wife Rohini have funded a not-for-profit initiative called EkStep, whose goal is 'to create a learner-centric technology-based platform to improve applied literacy and numeracy for 200 million-plus children in five years'.

Children learn better when their lessons are engaging, interactive and fun; to this end, EkStep plans to gamify the basic concepts of literacy and numeracy, turning them into engaging content that can be offered on smart devices like tablets or smartphones. 'Self-learning through gamification' is the mantra here—children can assimilate knowledge at their own pace, in a format which is simple and entertaining. This kind of learning can take place both in schools and outside them, allowing students to consume 'bytes' of individualized learning, much like the Khan Academy model.

The child-friendly EkStep user interface will be supported by collaborative content and a scalable technology platform, which can be distributed through multiple channels. Ravi Gururaj, the chairman of NASSCOM's product council, has said that a project like this 'leapfrogs the status quo, leverages technology to the hilt, delivers massive platform value and transforms early education across the nation for all classes of citizens'.[11] The fundamental idea behind EkStep has also received validation from Bill Gates, who thinks that, 'Rapid advances in education software on mobile phones will change the way students and teachers around the world learn every day.'[12]

When it comes to higher education, the traditional university experience and method of education are being challenged by MOOCs (Massive Open Online Courses). Starting when prestigious universities like Harvard, Stanford and MIT decided to make videotaped lectures available on the internet, today companies like Coursera, Udacity and edX offer entire courses online from some of the world's best universities, allowing people to sign up, view lectures and submit homework from anywhere in the world, creating a global education network. MOOCs have become hugely popular in India so much so that Indians make up the largest foreign student user base at Coursera, and are among the top users at Udacity and edX as well.[13] Students now have free access to world-class instruction from brand-name schools; they can also opt to pay for certain services, such as online exams or certificates of completion, which are used to bolster their resumes.

Some of India's top universities, such as the IITs, are part of these MOOC platforms, but we do have our own indigenous online education systems as well. The Government of India runs the National Mission on Education through Information and Communication Technology (NMEICT) which has launched an online education portal called Sakshat. In addition, the NMEICT is also in charge of the National Programme on Technology Enhanced Learning (NPTEL), in which seven IITs and the Indian Institute of Science are collaborating to create online content for science and engineering students.[14] The challenge now lies in integrating these online platforms into the regular curriculum. MOOCs are currently viewed as an addendum to traditional classroom-based coursework; the latter enables you to get a degree from a deemed university, while the former is more important for acquiring the kind of skills that will help you land a job, a crucial aspect in a country whose standards of higher education are so low that nearly half the graduates it churns out every year are unemployable in any sector.[15]

Granting choice, eliminating fraud

While MOOCs bring up regulatory concerns around innovation and the role of the state in education, there is another disruption that the government can bring about through incentive design. The concept of school vouchers has been discussed at length over the years, and was mentioned in *Imagining India* as well. This idea grants the power of choice to the consumer, and funds students instead of schools. Rather than pouring money only into government schools, a fraction of the funds can be used to grant school vouchers, which students can then use to pay for their education at a school of their choice. Naturally, students will gravitate to the institution that provides the best level of instruction, whether it's public or private; the voucher system brings in competition that can help to lift the overall quality of both sets of institutions. The flow of money from the government to schools will now follow the principles of the open market—the best-performing schools will get more money as more students enrol, and the underperformers will either have to pull up their socks or go out of business. Nandan envisioned the impact of school vouchers on the educational system as a reform which 'effectively removes ideology from funding and implementation and makes it easier, say, to hand over management of existing and failing government schools to the private sector, if this will attract students. This can bring the private sector and NGOs into already existing school infrastructure and government school buildings, instead of the current approach where we are constructing an alternative, private school system from scratch.' Pilot projects have been implemented by some of India's state governments, as well as by Parth Shah and his team at the Centre for Civil Society.

Aadhaar can be used in the creation of a central registry and voucher-issuance platform for schools and students. Students can be registered in the system using their Aadhaar numbers. Vouchers can then be issued against the Aadhaar number of the students, and parents can use these to enrol their children at a school of their choice. Section 12 of the RTE mandates that private and unaided schools set aside 25 per cent of their total enrolment capacity for students from economically disadvantaged backgrounds, and that such students should be given free and compulsory education up to the elementary level. This provision of the RTE Act gives us an opportunity to create a voucher system, where poor families are issued school vouchers that can be used by parents to match children to schools in the same way that students are matched to engineering or medical colleges upon completion of standard twelve.[16]

Lastly, we would like to turn our attention to the question of de-materialization of degrees and skill certificates. Prime Minister Narendra Modi has already announced a Digital Locker initiative in which a person's important records, including educational certificates, will be stored securely in the cloud and can be accessed by government departments as needed.[17] Fake resumes are circulating in the job market to an alarming degree, with an estimated one in five resumes in the IT industry being falsified.[18] People go so far as to set up fake companies that can provide experience certificates to jobseekers, helping them to inflate their expertise and skills when job-hunting.[19] A de-materialized degree combined with Aadhaar-based identification serves as a guarantee for the person's educational qualifications, increasing trust between jobseekers and potential employers. Individuals no longer have to get copies of their degree certificates attested and notarized, removing the friction generated by the need to authenticate paper copies of documents.

Another immediate application of Aadhaar in the educational sector arises in the wake of fraud during entrance examinations, typified by the long-running Vyapam scam in the state of Madhya Pradesh. 'Vyapam' is a Hindi abbreviation for the Madhya Pradesh Professional Examination Board, an autonomous body responsible for conducting

1 The Right to Education (RTE) Act is a landmark legislation.

2 There is a need for strong outcome-based measurement.

TEACHING THE NEXT GENERATION

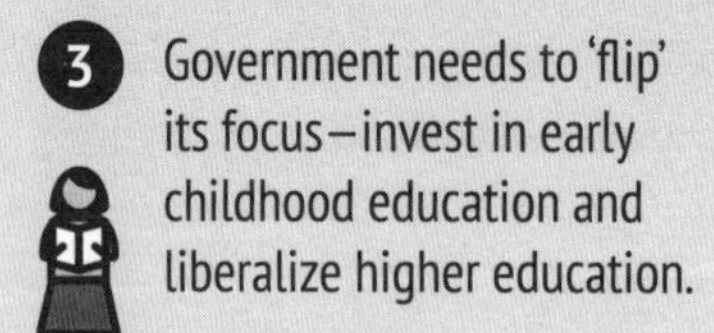

3 Government needs to 'flip' its focus—invest in early childhood education and liberalize higher education.

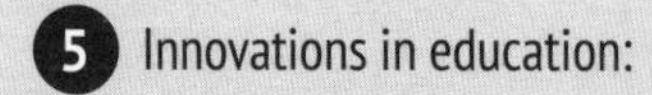

4 Flipped classrooms where students watch lectures at home and do homework in class.

5 Innovations in education:

A Course centric: Online education through Khan Academy and MOOCs.

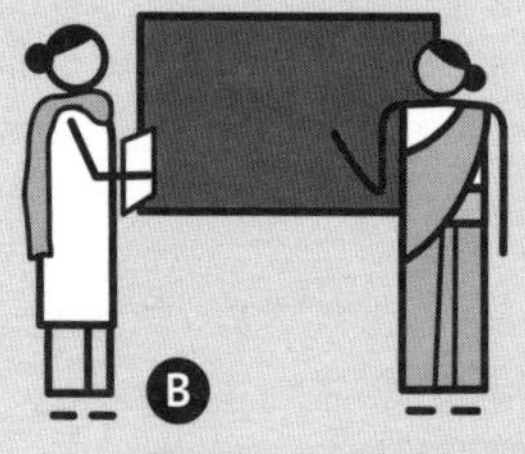

B Teacher centric: Better training.

C Learner centric: Using technology and gamification to improve learning.

6 An electronic national education network to manage all engagements between the government and students.

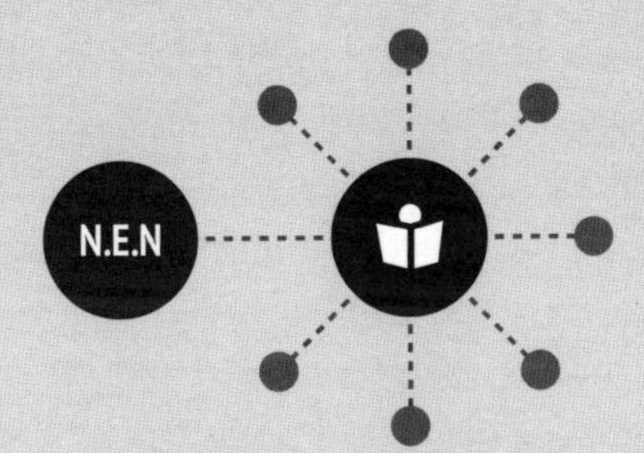

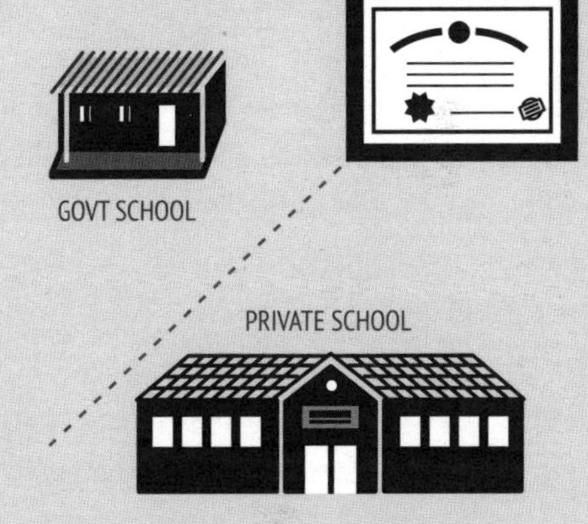

7 Software platforms that allow schools and students to be matched in keeping with RTE norms.

8 Changes in education policy to cover accreditation and quality control for private schools.

9 Education will become a continuous, lifelong process using new learning models.

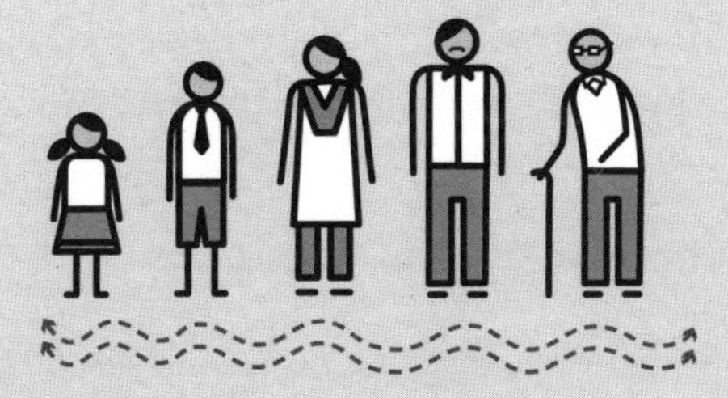

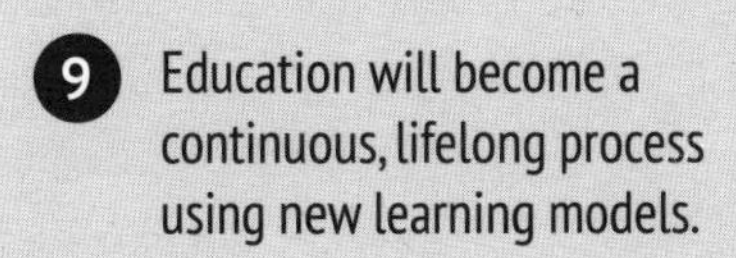

10 Government moves from being a funder to an enabler, providing regulation and integrating public and private innovations in education.

various entrance examinations in the state. Undeserving students colluded with unscrupulous middlemen, government officials and politicians to obtain high ranks in these tests in exchange for kickbacks; one of the most common methods of cheating was what Nandan refers to as the 'engine-and-bogey' scam. A good student would sit in the middle—the 'engine'—and on either side were the others—the 'bogeys' who would copy from him. How would Aadhaar help to eradicate such fraud? Nandan explains, 'When you have an exam, you authenticate (students) using Aadhaar to ensure they are genuine candidates. And you don't issue a seat till they enter the room and then you randomize and give them a seat. This is not rocket science, it's simple. And people are getting killed on this.'[20]

The de-materialization of educational certificates as an actionable idea was first suggested in 2010; a National Academic Depository Bill was tabled in Parliament in 2011, and the current administration is expected to take this initiative forward.[21] Just as stock depositories store share certificates in electronic form and have a regulator to oversee their functioning, so also a National Academic Depository will store electronic certificates; the role of the regulator in this case would be to verify the data contained in these certificates for accuracy. Similar organizations already exist; a subsidiary of the NSDL, together with NASSCOM, operates the National Skills Registry, designed to provide potential employers with verified information about jobseekers in the IT industry. Such a depository would make document-sharing easy while also guaranteeing the authenticity of the information it stores.

Changing the way India learns

In the preceding diagram, we lay out a roadmap for the transformation of India's education sector. New laws have ushered in an era of reform in education, and new innovations are flipping the way students learn in and out of the classroom. We must use these developments to address some of the most crucial problems bedevilling our education system. Since early childhood is the critical phase for learning, we advocate that the government 'flip' its focus, concentrating on early education and liberalizing the higher education segment. Right now, the government is the highest authority when it comes to dictating which course must be taught at which level. What if the government took the radically different approach of letting institutions decide their own study plan for their students? In such a case, the government can create a user-rating system, along the lines of those used by services like Uber and eBay, to drive trust in educational providers. It is now the users who decide the rating and quality of a provider, rather than the bureaucracy.

An NIU can be created to simplify the engagement between government and students, managing, for example, the voucher system under the RTE Act to match schools and students. This NIU must also operate with the idea that education is no longer restricted to the standard classroom model; education should become a fluid, lifelong process, whether a student is learning from a teacher or from a game on a smartphone. To accommodate this transition, the government should move from being a funder to an enabler, creating the regulatory and policy environment that allows new educational models to flourish.

Wherever you look in India's education system, you are confronted with issues that need urgent resolution. Millions of children never get any kind of formal schooling at all. Of those that do, the education they receive is often substandard. The government is pouring money into a public school system that is utterly inadequate to meet the challenges of educating India's poorest and most deprived children; parents are turning in droves to the private education sector at great financial cost, even though the quality of education offered by such institutions can vary. Our institutions of higher learning are churning out people with degrees who have no skills whatsoever, rendering them unfit for employment. Underqualified individuals are resorting to fraud to get a job out of sheer desperation.

There is no magic bullet that can cure education in India of all that ails it. The lack of a good education closes doors of opportunity for people, consigning them to the fringes of economic growth. The lost potential, both for them and for the nation, is immense and heartbreaking. Just as in the case of Aadhaar, such a fundamental social problem needs a concerted effort across the spectrum of organizations—government, the private sector, NGOs and others. We need to implement innovative new ideas—many of them—and most of all we need to recognize the value that technology-based innovations can hold, not just to change the way students learn, but to radically alter our educational landscape.

14

SWITCHING ON OUR POWER SECTOR

> We will make electricity so cheap that only the rich will burn candles.
>
> —Thomas Alva Edison

LOHA SINGH MAKES for an unlikely film star, but this weathered, obscenity-spewing resident of Kanpur, located in India's Hindi heartland, is the hero of the award-winning documentary *Katiyabaaz*. The title refers to his job as the Robin Hood of power, illegally rerouting electric wires from affluent areas to localities without electricity. In a city where the average summer temperature is a sweltering 48 degrees Celsius and neighbourhoods are plagued by power cuts that last up to fifteen hours a day, residents are willing to pay this crack electricity thief—he handles live wires with his bare hands, and has even blown up a transformer so that he can reroute the wires inside—to ensure an uninterrupted power supply. In the meantime, the Kanpur Electricity Supply Company Limited (KESCO) is reeling under heavy losses thanks to increased demand, inadequate supply, unpaid bills, and of course, the thefts that Loha Singh and his ilk perpetrate. While *Katiyabaaz* is an enjoyable, thought-provoking watch, it's also a primer to some of the biggest issues plaguing India's energy sector.[1]

Precariously poised on the cusp of climate change

Energy resource management is such an essential component of India's economic growth that Nandan devoted an entire chapter in *Imagining India* to understanding the origin and impact of India's energy policies and evaluating the state of the field at the time. Since then, India has grown to become the world's fourth-largest energy consumer, heavily reliant on foreign imports to meet its energy requirements and likely to become one of the most import-dependent countries in the world.[2] Our energy economy continues to be largely dependent on fossil fuels—India is the world's third-largest consumer of coal, for example, with all the attendant implications for environmental degradation, global warming and climate change.[3] Heavy subsidies for the conventional energy sector discourage large-scale innovation into alternative energy methods, while draining the pockets of the state-run electricity distribution boards, leaving them unable to pay for the power consumers need. As we have mentioned earlier, global oil prices have been falling over the past few years, making this the perfect time to seriously rethink and re-architect India's energy policies.

Inefficiencies in electricity distribution are grossly distorting the balance between energy production and consumption. Power losses due to dissipation from electric wires or theft, estimated as the Aggregate Technical and Commercial (AT&C) losses, are second only to China and the US. Dissipation losses arise due to the physical nature of the transmission network; they are largely unavoidable, and tend to be the same worldwide. It's the theft component that is throwing off our numbers. When such losses are stacked up against the energy these three countries consume, it is clear that China and the US are operating at relatively low inefficiencies, whereas a staggering 30 per cent of India's power is being siphoned off by the illegal tapping of power lines and the bypassing of electricity meters, often with the connivance of electricity board employees.[4] Systems for metering electricity usage, billing and revenue collection are also weak and lead to further revenue losses for the public sector distribution boards. Some 300 million Indians continue to live without electricity, and 43 per cent of rural India depends on kerosene for lighting up their homes.[5] According to World Bank estimates, annual losses from India's power sector could reach $27 billion as early as 2017.[6]

The alarming increase in carbon dioxide emissions and climate change are now global concerns, providing an impetus for the development of alternative energy sources. While we may argue about who is to blame, the fact remains that tropical regions such as the Indian subcontinent and Africa stand to become the worst hit unless immediate and urgent action is taken to invest in renewable energy sources. In this regard, India's lack of a legacy infrastructure and low adoption of technology is actually an advantage; we are free to build the institutions that best meet the needs of our rapidly changing energy sector, and the successes and failures of other countries in the adoption of green energy technologies can act as a road map to guide us forward.

The energy grids of our future will look very different from the kind of energy grid that distributes electricity in India today. Instead of depending only on traditional modes of power generation, like coal, gas and to some extent nuclear power plants, all of which have a steady output, energy systems will also need to integrate intermittent energy produced from renewable resources such as hydroelectric power plants, wind turbines and solar panels, as well as from biofuels, geothermal, tidal and ocean thermal sources. 85 per cent of India's rural households continue to depend on biofuels (firewood, cow dung) as their primary energy source, and these have been completely off the grid so far.[7] They also bring with them a more insidious problem—the fumes generated by burning biofuels indoors can cause respiratory problems so much so that, according to the World Health Organization, India has the world's highest rate of death due to chronic respiratory disease.

In the future, some consumers could also become generators, for instance, feeding excess power generated by a solar panel back into the system. Conventional transmission and distribution grids cannot handle the complexities of intermittent supply and reverse energy flow efficiently. The grids in operation today function as a one-way street, with both electricity and information flowing from the power company to the consumer. The consumer, on the other hand, has little to no idea of how much their power consumption is until provided with a bill, nor any idea of which type of energy source—renewable or non-renewable—was used to generate the power they consume; the only power the consumer has over the system is to notify the utility in case of a power cut.

Greening India's power sector: Smart grids and renewable energy

Increasing the efficiency and transparency of our power grids is possible only when they become smart grids, incorporating the power of technology and digital processes to change the energy landscape of our country. Such smart grids are two-way channels of communication, allowing energy utilities to integrate renewable energy sources into the system and monitor their networks more effectively while providing more information to consumers about their energy usage. By promoting the use of alternative energy sources as well as the more efficient usage of fossil fuels, smart grids can cut down the emission of carbon dioxide and other greenhouse gases and help to rein in the runaway problem of climate change. From Italy to the USA, countries around the world are embracing the smart-grid model with the dual goals of increasing operational efficiency and reducing the impact on the environment.[8] While comprehensive data is not available, a 2011 study by the Electric Power Research Institute (EPRI) suggests that while the net investment needed to realize the vision of a smart grid will be nearly $500 million (over a twenty-year period), the benefits from such a power system will run into trillions, with a benefit-to-cost ratio of 2.8 to 6.[9]

What goes into the building of a smart grid? For power companies, sensors installed on power lines provide real-time information on the health of the network, allowing utilities to detect loads, congestion and shortfall, and marshal their resources more effectively to ensure a smooth and uninterrupted power supply. Faults can be repaired remotely, obviating the need for a technician to manually do the job. Such sensors can also act as an early warning system in case of a power outage so that the problem can be rapidly identified and fixed remotely before it snowballs into a massive blackout like the ones that hit north India in 2012—the largest in recorded history.[10]

For consumers, smart meters can track electricity usage, transmitting information back to the power companies and to the consumers themselves. Consumers know exactly how much electricity they have been using, and this information can help prevent billing disputes. Utility companies can also start implementing time-of-day pricing schemes, charging customers a higher rate during times of peak load on the system. Pilot studies show that providing usage data to homeowners results in an average drop of 3–5 per cent in household electricity consumption.[11] By allowing utilities to record usage patterns, smart meters also make it easier to detect theft and misuse. Italy has been a pioneer in the smart meter field; over 30 million smart meters have been brought into service since 2001, and 85 per cent of all Italian households now use smart meters to manage their electricity.[12]

The ability to monitor power sources in real time allows utilities to respond to demand-and-supply forces rapidly and with much greater accuracy. This makes it possible to integrate smaller and intermittent sources of power, such as wind turbines and rooftop solar panels, into the power supply system. In the future, smart grids can also accommodate the draw on energy by electric cars being recharged. Power utilities can use smart grids to improve their operational efficiency for maximal utilization of existing energy sources, as well as the integration of renewable energy sources into the system. Energy efficiency has in fact been dubbed the 'fifth fuel', and Amory Lovins of the Rocky Mountain Institute has coined the term 'negawatt' to describe power saved through efficiency or conservation.[13]

Energy efficiency is being driven by innovations across multiple areas.[14] Renewable energy sources, in particular solar energy, have boosted the available energy supply, and consumers can now act as small producers and storers, in effect 'decentralizing' the power

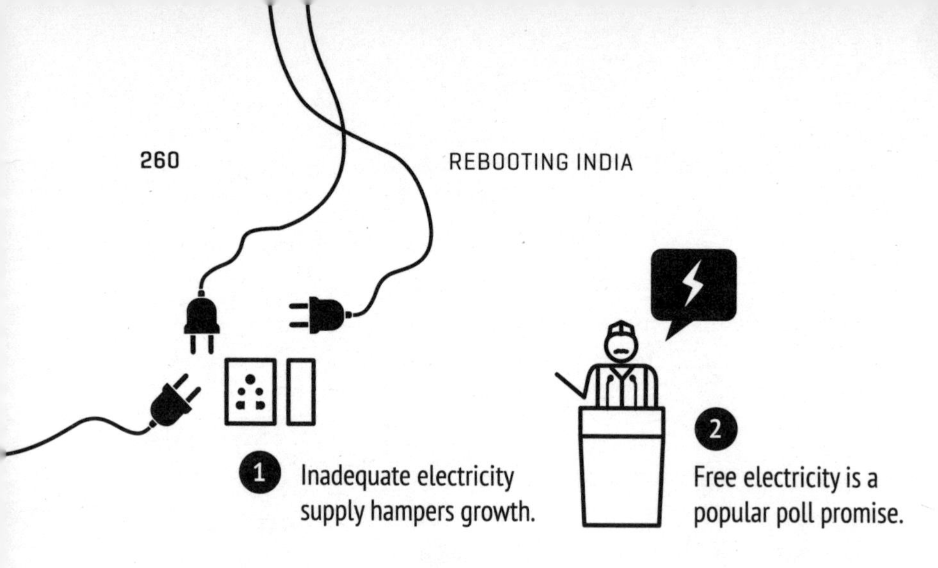

1. Inadequate electricity supply hampers growth.

2. Free electricity is a popular poll promise.

ROAD TO A HEALTHY POWER GRID

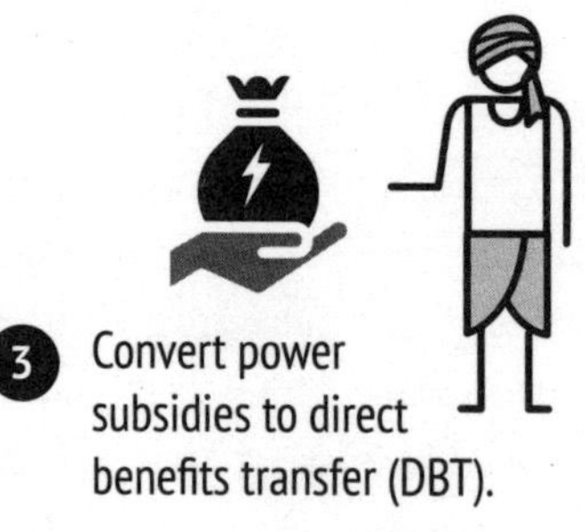

3. Convert power subsidies to direct benefits transfer (DBT).

4. DBT removes the burden of subsidy administration from state electricity boards. They can focus on their core job and generate profits.

5. Power grids will become complex with renewables (e.g. wind) and intermittent producers (e.g. rooftop solar panels, electric cars).

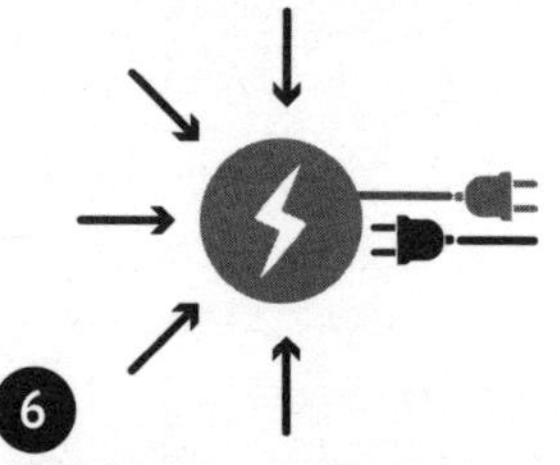

6. Power grids must become two-way and develop additional storage capacity—a 'smart grid'.

7

Stop building capacity for peak power; store energy in newer, cheaper batteries.

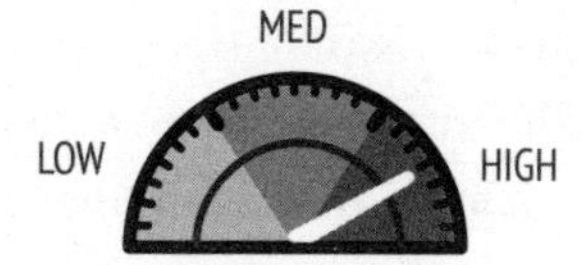

8

Dynamic pricing that allows customers to optimize power consumption for smart energy management.

9

Feed-in tariffs as incentive: Consumers who feed power into the grid through solar panels, wind-powered generators or electric vehicles get paid.

10

Carbon credits will prompt people to offset their CO_2 emissions.

Building new institutions to handle energy management: the role of the government

1 Setting up a National Energy Network to manage energy subsidies—direct benefit transfers to consumers, and subsidies to renewable energy producers.

2 Creating common standards and protocols for the millions of producers, distributors and consumers who will be linked to each other in the grid of the future.

grid.[15] Storage is getting cheaper—the batteries that power Tesla's electric cars may soon be made available for the home as well.[16] Smart systems are managing power consumption more efficiently. All of these factors are leading a sweeping transformation of the traditional energy grid structure, with resultant improvements in performance and savings. The importance of bringing about similar changes in India was commented upon by former union minister Jairam Ramesh, who emphasized that 'chasing gigawatts alone is not enough and will not be sufficient for true energy transformation. What is required is a simultaneous push to ensure decentralized generation and distribution as well so that local communities who do not have access to electricity now get power in their homes.'[17]

India's energy production patterns vary dramatically from state to state. Today, Gujarat can boast of an energy surplus, while Andhra Pradesh is struggling with a 12.1 per cent energy deficit. Smart grids can help to level the playing field, by drawing surplus power into the system and distributing it where needed in a planned manner, irrespective of the physical location of the power plant and the type of energy it's using to generate electricity. Smart grids can also be used to bring India's vast unmetered population, largely agricultural, on board, as well as help to identify which consumers are eligible for subsidized electricity and preventing fraud.

Turning on the lights: The road to a healthy power grid

The preceding diagram lays out a road map towards building a smart power grid in India. Much of the groundwork for these ideas is contained in two reports, 'IT Task Force Report for Power Sector' (2002) and 'Technology: Enabling the Transformation of Power Distribution in India' (2008), an updated version of the 2002 report that was prepared by Infosys and the Centre for Study of Science, Technology and Policy (CSTEP).[18] Nandan chaired the committee that prepared both reports, and, seven years later, they still offer a critical perspective on what technology can and cannot do for India's power sector:

> It can offer a framework for an efficient power system, providing the technical design of a future smart grid and help us address commercial and behavioural issues. In the technical design of a smart grid, IT can help monitor and control electricity real-time with fine granularity, construct a robust, self-healing grid—detect outages, load, congestion and shortfall, and establish two-way power exchange with a large number of renewable generators, storage devices and devices such as plug-in hybrid vehicles. In terms of commercial and behavioural issues, IT can help identify theft and losses, provide choice to customers, allow for new pricing mechanisms such as Time-of-Day (ToD) or real-time, enable much improved transparency and conservation, and provide the structure for sophisticated billing, collection and information management.

So, what are technology's limitations in this sphere? The report states:

> We must remember that IT by itself cannot change practices and fundamentals directly. This includes altering the fundamentals of the power system—including supply and demand, tariff structure and power quality—changing operational practices and organizational culture, reducing losses and theft, meeting environmental challenges, and changing governance and project management. People, especially trained and equipped personnel, are the critical players in affecting this transformation.

Any such solution needs to take into account the crippling burden that power subsidies place on the administration. We advocate a move towards a direct benefit transfer model for subsidies so that state electricity boards can focus on their key mandate, that of providing reliable power to all citizens, instead of diverting their energies towards subsidy management.

Renewable energy sources may still be in the early stages of growth in India, but they hold tremendous potential. The sight of solar panels dotting the roofs of city houses is no longer unfamiliar, and the new office of the ministry of environment and forests in New Delhi is powered by a solar rooftop system. As these intermittent power sources start plugging into the grid, they will need to be handled efficiently. Small producers will start contributing to the network, giving rise to two-way energy flows. Grids will need to develop the storage capacity to handle such inputs and integrate them into the management of demand and supply. Instead of building power generation plants to a peak power capacity, the declining price of storage can be leveraged to build low-cost battery systems that can store power as needed and supply it in the face of high demand. A cautionary tale in this regard comes from Hawaii, where the supply of solar energy from homes has so overwhelmed local power utilities that, until recently, customers were banned from installing new solar rooftop systems; power companies there and elsewhere are trying to curtail the rise of solar energy by making it financially unviable for both customer and solar energy companies.[19]

We also anticipate a change in the behaviour of the average Indian consumer, from being a passive recipient of power to an active participant in the power grids of the future. They can optimize their power consumption patterns through the use of innovations like Nest that provide real-time information on power usage. Feed-in tariffs—where customers get paid for supplying energy to the grid, generated by the use of electric cars or rooftop solar panels, for example—can incentivize the use of renewable energy sources, and the concept of clean energy and carbon credits can be brought to consumers to promote the widespread adoption of energy-saving practices.

When it comes to the organizational management of smart grids, India's crumbling public-sector energy utilities, beset by inefficiency, corruption and heavy financial losses, are inadequate to shoulder the burden. We need a mechanism where we can draw upon the expertise of those from the government as well as from the energy sector, and bring in multiple stakeholders to function in unison—power utilities, regulators, policymakers, technology service providers, standards bodies, consumer groups and experts in the fields of technology, economy, law and others. Another key function of the government in creating a new model of energy management would be to promote innovation in the private sector by setting in place common standards and protocols that can serve as the basis for new operating models in the production, distribution and consumption of energy.

India's power sector is precariously poised today. We can no longer use subsidies to insulate ourselves from the shocks in worldwide energy pricing, we cannot turn a deaf ear to the increasing clamour around the use of fossil fuels, and we cannot ignore the fact that our energy policies and practices are completely unsustainable in the long term. We have an unparalleled opportunity to implement technology-powered solutions that can pave the way for India to become a pioneer in building smart, sustainable energy solutions that can power the nation's economy forward.

15

JUSTICE DELAYED IS JUSTICE DENIED

At his best, man is the noblest of all animals; separated from law and justice he is the worst.

—Aristotle

AT THE RISK of stating the extremely obvious, the Indian judicial system is in trouble. By any measure you choose, our judiciary is a woeful underperformer, whether it is in the number of judges and courts available to us, or the speedy disposal of cases placed before them. We have only ten to fifteen judges per million people, and our judge-to-population ratio is among the lowest in the world.[1] As of 2014, the total number of pending cases across India was a staggering 30 million.[2] If the judiciary at its existing strength were to roll up its sleeves and tackle the Herculean task of resolving all the cases pending across the country, Andhra Pradesh High Court Judge V.V. Rao estimated it would take them 320 years to clear the backlog.[3] A few years ago, Delhi Chief Justice A.P. Shah admitted that the Delhi High Court was lagging behind by 466 years; in an attempt to clear up its caseload, the court decreased the average time spent hearing a case to a little over five minutes.[4]

Even five minutes in court would be a welcome relief for the 300,000 inmates of India's prisons who languish in custody while waiting for their case to be heard. Their numbers are growing steadily, and many of them are either illiterate or poorly educated. If their case is to be heard by a higher court—as in the case of an appeal—they're in for another endless wait, since it takes an estimated fifteen to twenty years for a case to wend its way from the lower courts all the way up to the Supreme Court.[5] No wonder, legal cases in India often become family heirlooms, with the children and grandchildren of the litigants faithfully continuing to fight for their family's cause long after the original complainants are dead and gone.

Our ailing judicial system

While mind-boggling in scale, none of these problems are particularly new. Starting with the pre-Independence-era Rankin Committee in 1924, there has been an endless parade of committees, commissions, conferences and reports, all dedicated to the problem of judicial reform. Clearly there is no shortage of ideas when it comes to increasing judicial efficiency, including such obvious measures as hiring more judges and building more courts. Why haven't they been implemented?

Part of the answer is money. India spends less than 1 per cent of its GDP on the judiciary, far less than that in other democracies.[6] More than half of this paltry sum is raised by the judiciary itself through court fees and fines. Another is the set of checks and balances placed on the judiciary by the executive and the legislature. As Alexander Hamilton wrote in the Federalist Papers:

> The judiciary, on the contrary, has no influence over either the sword or the purse; no direction either of the strength or of the wealth of the society; and can take no active resolution whatever. It may truly be said to have neither FORCE nor WILL, but merely judgment; and must ultimately depend upon the aid of the executive arm even for the efficacy of its judgments.[7]

While meant to ensure an equitable distribution of power, these constraints do have a major impact on the functioning of the judiciary. So much so that in 2007, the then chief justice of India, K.G. Balakrishnan, laid the blame for the glacial pace of India's legal system squarely at the feet of the government, saying, 'The main cause for judicial delay lies not so much with the judiciary as with the executive and administrative wing of the government.'[8]

At this point in time, the problems of the Indian judiciary are too deeply entrenched to be solved by simply throwing more money at the system to appoint more judges or build more courts. In recent years, we have seen the emergence of fast-track tribunals in an attempt to speed up the process of dispensing justice, but even these are the equivalent of trying to stem a fire hose with a cork. We cannot think of any other aspect of Indian society where it is so clear that the standard solutions simply will not work. It's time to redesign the judiciary's operating system, and create a new class of processes and institutions that will make our legal system fast, fair and transparent. Technology is the foundation upon which these processes and institutions must be built.

Cleaning up our courts with technology

The first class of reforms includes the creation of technology-based systems to handle the everyday minutiae of the legal process. Nearly 90 per cent of the work involved in preparing a case for trial does not need a judge's intervention. This consists of such routine tasks as generating documents, collecting evidence and scheduling trials, while ensuring that the laws regulating these tasks are being followed. Such activities can easily be systematized and speeded up by developing technology-based applications to handle data and monitor progress, lessening the burden on judicial officials and saving precious courtroom time that is otherwise wasted on debating trivial matters of procedure.

Today, India's judicial system continues to remain largely opaque, with an ordinary individual having little information and even less control over the progress of their case through the courts. Procedures may often seem cumbersome and confusing, leading to their being followed incorrectly and adding to the delay. A centralized, technology-powered court operations platform would help to address these issues, and this is the second class of reform we propose.

Such a platform could function as an electronic repository of all case records and documents, allowing both judges and participants to search the database for relevant information. This becomes especially important in a system like ours, which is almost entirely paper-dependent and routinely generates reams of paper for each case. India's longest running trial, the 1993 Mumbai bomb blast case, generated over 120,000 pages of legal material, which had to be ferried to court in trucks.[9] Misplacing even a single sheet of paper can result in inordinate delays, and digging through a library's worth of paper to find a necessary document or piece of information is an enormous waste of time. Creating an electronic database makes it easy to keep track of judicial records, as well as setting in place standardized processes for the capture and storage of information, thereby reducing ambiguity in the system and eliminating delays due to incorrect record-keeping.

A court calendar with details of court activities and hearings will make it extremely easy to follow the progress of a case, and will also allow for metrics to be generated that measure efficiency and flag inordinate delays as requiring further investigation. Cases are often prolonged indefinitely thanks to an endless series of adjournments; a 2000 amendment to the Civil Procedural Code sought to address this issue by allowing only three adjournments per case. The court calendar can be used to track the number of adjournments granted, and cases that have reached the three-time limit can be flagged for immediate hearing, ensuring that a major reason for legal delay is eliminated instantaneously.

Each individual case can be monitored via a dedicated case planner, where judges define a plan for each case based on its category or profile. Plans can be tracked and rescheduled on an ongoing basis depending upon case progress, and a running schedule can be maintained for all hearings and events regarding the case. The status of each case can be monitored via a case dashboard, which allows for easy detection of delays, and generates overall performance data for a particular court with respect to progress and disposal of cases. E-filing systems can also be created to allow many judicial processes to be completed digitally rather than requiring the physical presence of the parties involved.

A system of this nature is designed to be inclusive, drawing in judges, litigants, advocates, police, jail authorities and citizens to allow for smoother functioning and to correct inequalities in information distribution. It will also allow for systematic case planning, help to save time by more efficient planning of court hearings, reduce the time spent in looking for case-related material, aid in the decision-making process, improve accountability and decrease delays across multiple departments and institutions, and most importantly, make the justice system more citizen-friendly.

An institution for change: The National Judicial Network

Technology-based judicial reform efforts have been undertaken in India, but their implementation continues to be limited at best. For example, the chief justice of India had called for the creation of a planning and management system for the judiciary; in response, the National Judicial Academy presented to a group of judges in Bhopal in 2008 a plan for what they called an 'Information Management System for Administration of Justice' (IMAJ), a system that would function much like the judicial platform we have discussed earlier; this proposal did not move beyond the planning stage. In 2011, the government approved the creation of the National Mission for Justice Delivery and Legal Reforms, envisioned as a sweeping overhaul designed to render the judicial process both fast and fair.[10] An e-courts project has also been established to help digitize judicial processes,[11] but the ambitious goals outlined by such initiatives still remain distant. In 2008, Nandan himself had met with Dr Mohan Gopal, then the director of the National Judicial Academy, as well as with a Supreme Court judge to discuss the implementation of a centralized courts management system, but the initiative did not take off.

Governments around the world are creating new types of institutions that allow their judicial systems to function more efficiently. In the United States, for example, the Administrative Office (AO) of the United States Courts was set up as far back as 1939 to handle all the support functions that are part of the legal process. The AO now provides lawyers, public administrators, accountants, systems engineers, analysts, statisticians and other staff as needed—in effect, the routine tasks needed for frictionless functioning of the courts have been outsourced to such organizations (although they are still part of the government), leaving the courts free to concentrate on their fundamental purpose of delivering justice.[12]

India desperately needs the services of a similar organization to reduce the inordinate burden crippling our courts. We propose a road map towards building a better judicial system through the creation of a centralized court operation platform, an entity that will drive the transformation towards completely paperless justice delivery. All documents and evidence will be digitized and easily searchable, and the progress of a given case through the courts can be monitored, allowing for better case tracking and performance management. Some of this data can be made available for public analysis so that trends can be mined to improve the overall functioning of courts. The centralized platform should also allow stakeholders to enrol at their own pace, phasing out paper-based systems in a controlled manner.

We envision that a National Information Utility—the National Judicial Network (NJN)—would be the most desirable organizational model to build and oversee such a platform. The judiciary guards its independence fiercely, and the structure and functioning of the NJN must completely protect that independence. For instance, the NJN can operate under the aegis of the Supreme Court, much like the National Judicial Academy currently does, and its funding can come entirely from the allocated court budgets. The stakeholders that have a say in the functioning of the NJN should be those institutions that would be the prime users of its services, namely the law ministry, the Supreme Court and the High Courts.

To effectively handle its mandate, the NJN would need the services of both legal experts and technologists, each with a working knowledge of the other's domain so that processes can be simplified while still adhering to the letter of the law. Today, judges have to preside over

1 In 2014, there were **3 crore** pending cases across India.

JUSTICE DELAYED IS JUSTICE DENIED

A Road Map to Timely Justice

2 We spend very little on our judicial system.

3 Courts waste time on procedure, not justice.

4 Technology to streamline our courts through a National Judicial Network (NJN).

5 NJN funding to come from court budgets to retain independence.

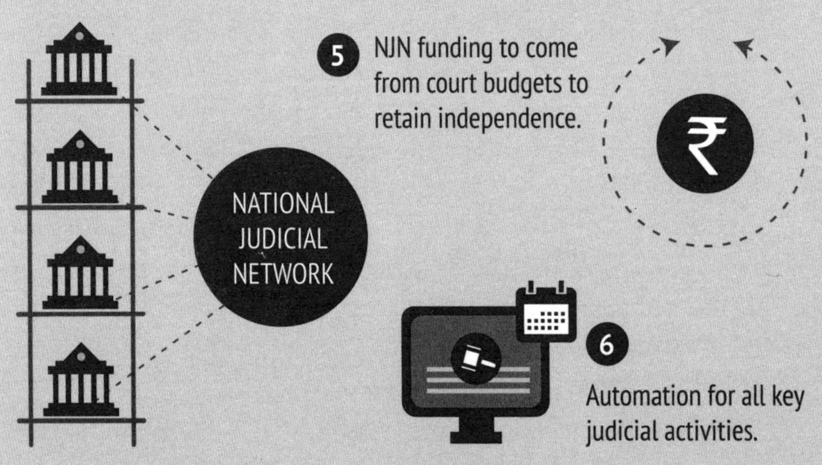

6 Automation for all key judicial activities.

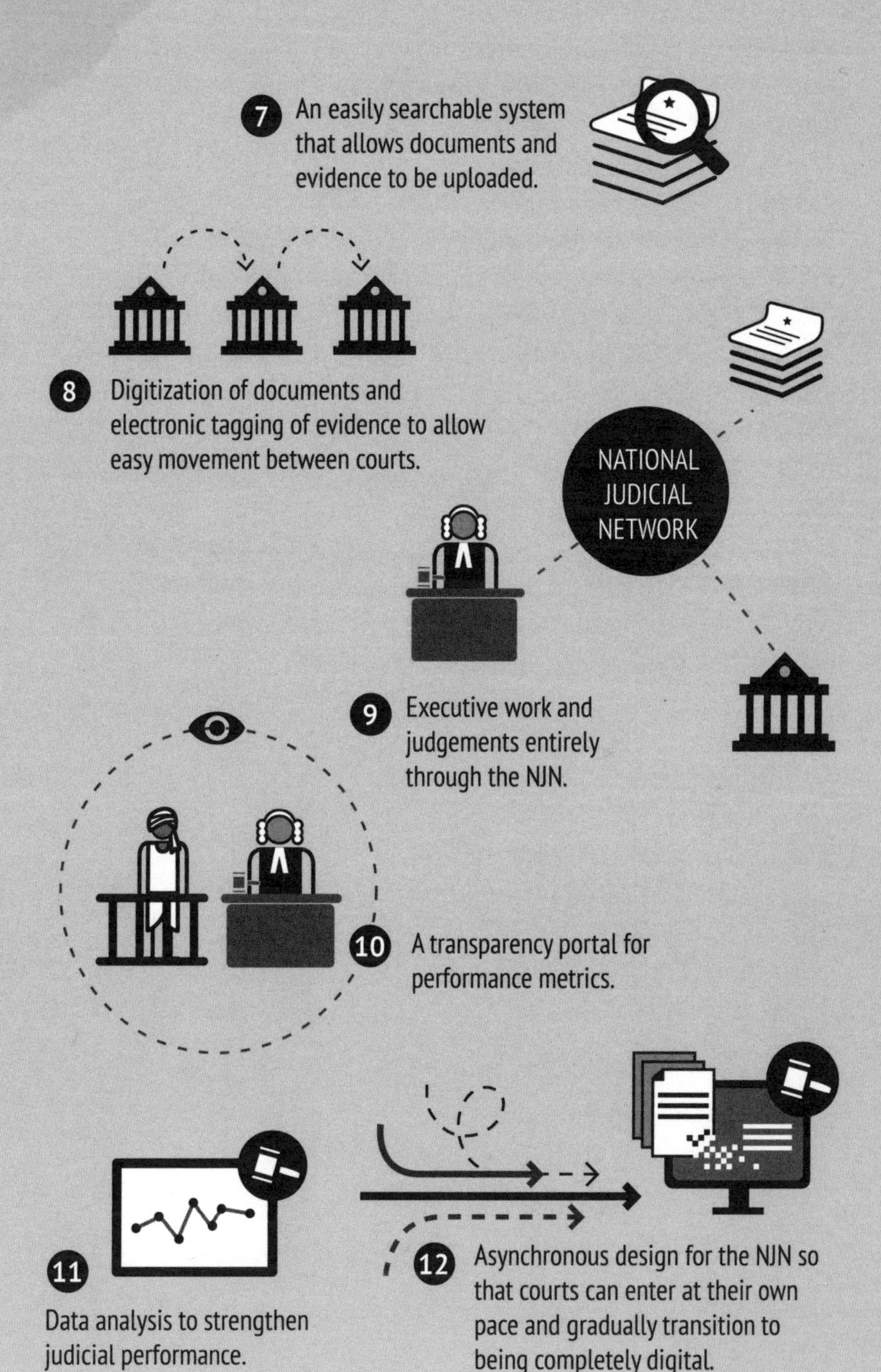
7 An easily searchable system that allows documents and evidence to be uploaded.
8 Digitization of documents and electronic tagging of evidence to allow easy movement between courts.
NATIONAL JUDICIAL NETWORK
9 Executive work and judgements entirely through the NJN.
10 A transparency portal for performance metrics.
11 Data analysis to strengthen judicial performance.
12 Asynchronous design for the NJN so that courts can enter at their own pace and gradually transition to being completely digital.

complex cases involving environmental disputes and patent litigation, which cover multiple fields of scientific and legal expertise. Rather than expecting judges to educate themselves in all these fields, the NJN should recruit experts in these areas, who can bring their diversity of training and experience to bear on the judicial process, reducing the burden of knowledge on a single individual in the system.

The objective of law is to create a balance between the citizen and the state. It protects the minorities from the tyranny of the majority. At the same time, all citizens must be held equal before the law. A situation in which the poor and underprivileged are trapped in the serpentine coils of an opaque and glacially slow-moving judicial system violates every principle of due process and equality. If we are to redress this wrong, we must design new institutions to strengthen our judiciary and infuse it with new talent that can promote more efficient and transparent functioning. Only then can we fulfil the Aristotelian definition of a stable state—one in which 'all men are equal before the law'.

16

REBOOTING INDIA: REALIZING A BILLION ASPIRATIONS

The future is already here—it's just not evenly distributed.

—William Gibson

India's twelve grand challenges

THROUGH THE COURSE of this book, we have discussed twelve grand challenges—and the platforms to address them—that directly affect citizens, businesses and the government. In the new innovation economy, the government has to think in terms of building platforms and ecosystems, developing minimal and simple solutions that can scale up, and providing a reasonable regulatory regime.

The first two grand challenges that have been tackled are Aadhaar and PaHal (the LPG subsidy). Today, there are over 900 million Aadhaar holders, and 200 million of these numbers are directly linked to bank accounts. Almost 140 million LPG consumers receive their subsidy in the form of a direct transfer into their bank accounts, a system that substantially decreases diversion through fraud and lessens the government's subsidy burden. To date, $3.8B (Rs 250 billion) has been transferred as LPG subsidy to consumers. Both these platforms are now firmly established. It is easy to implement the same system for a variety of other subsidy and social security programmes.

The next four grand challenges that we have described are those where the design is in place and early implementation is in progress, but scale has yet to be achieved. We have discussed the roadmap to a cashless economy with electronic payments. MicroATMs will bring banking services to every doorstep in the country. With e-KYC, we will begin the march towards a paperless society, where a range of services, from opening a bank account to availing of a government benefit, can be carried out electronically—no more waiting in line for hours clutching reams of paperwork. Eventually, going paperless and streamlining cumbersome processes will make our government and businesses more accessible, and our economy more inclusive.

Electronic tolls, the architecture for which is now being implemented, will give every vehicle a unique identity, its own Aadhaar. Trucks and people will now move effortlessly across our highways, cutting down on wait times at toll plazas. When implemented, the Goods and Services Tax will create a single common market in India, and through fraud prevention will increase government revenues over time. The data generated by the GST will contain important information about a merchant's commercial activities, and can be used by lenders to assess creditworthiness and provide loans to millions of small merchants. In fact, the data generated by GSTN can be combined with data from various other e-commerce platforms, payment systems, and MCA21—a platform launched by the ministry of corporate affairs—to make a data-based credit assessment of a business. Data can also become the basis by which individuals are granted loans, thanks to new credit-scoring models that combine information from payment transactions, tax filings, social media activity, and so on. In the same way that a computer algorithm can end up giving us a self-driving car, so also algorithms can unlock credit for millions of people and small businesses.

Finally, in the last six grand challenges, we extrapolate our learnings and experiences from the previous six. The use of technology already ensures free and fair elections in India today. As the Election Commission undertakes its stated goal of linking voter

CONSUMER DATA

ENTERPRISE DATA

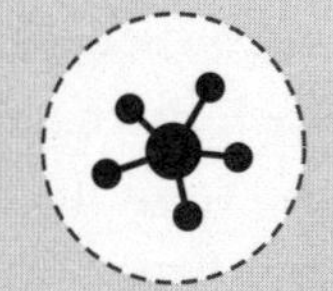

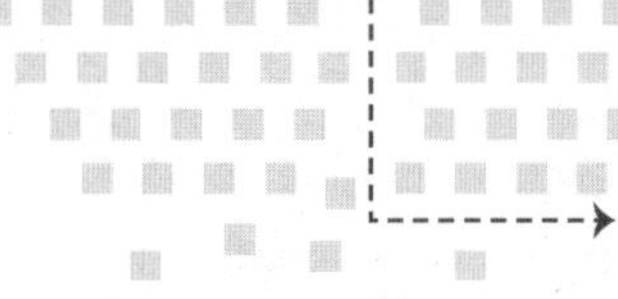

CREDIT IN A DATA-RICH NATION

CREDIT WITHOUT FRICTION WILL BRING MILLIONS INTO THE FORMAL ECONOMY

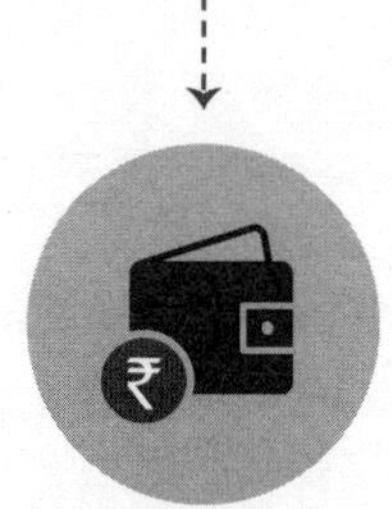

IDs to Aadhaar, the errors of inclusion and exclusion that plague our voter rolls and weaken our democracy can be fixed. We have also laid down a road map for technology-based interventions to rethink our educational system, switching focus to outcomes rather than inputs. In the healthcare space, we outline a plan for the nationwide implementation of electronic health records, designed to keep pace with new innovations in the area, such as the onset of wearable devices that can function as diagnostic tools. In the energy sector, we discuss how subsidies can be streamlined, as well as the need for a next-generation smart grid to cope with a world where millions of small producers start generating energy from their electric cars or rooftop solar panels. We identify the areas where technology can be used to overhaul our crumbling judicial system. Finally, we believe that the only answer to combating graft is a systemic one, where every rupee of government spending is tracked through an open system—the Expenditure Information Network.

Together, these twelve platforms will make it easy for our people to be well educated, healthy and productive, for our businesses to flourish, for our government to collect adequate revenues, and to operate social security programmes efficiently so that the less fortunate do not slip through the cracks. Robert F. Kennedy famously said:

> The gross national product does not allow for the health of our children, the quality of their education, or the joy of their play. It does not include the beauty of our poetry or the strength of our marriages; the intelligence of our public debate or the integrity of our public officials. It measures neither our wit nor our courage; neither our wisdom nor our learning; neither our compassion nor our devotion to our country; it measures everything, in short, except that which makes life worthwhile.[1]

India is poised to become one of the three largest economies in the world over the next few decades, with the potential to become a global superpower.[2] But this will not happen if we continue to tread well-worn paths and allow the benefits of growth to remain concentrated among a privileged few. We need technology-fuelled structural reforms in our government, our businesses and our society at large.

What is government?

While we speak of radically transforming government, let's also take a step back to ask a fundamental question: What is government, and what is its purpose? This is a question that we ourselves have mulled over during our tenure in government. Depending on who you ask, there are a multitude of answers. Some describe government as a collection of institutions—the legislature, the executive and the judiciary, governed by checks and balances that regulate the separation of powers. Some describe government through the lens of the economy, labelling it socialist, capitalist, and so on. For some, government is described by its political inclinations, left wing or right wing. For the common man, the government is the local politician and the local bureaucrats with whom he regularly interacts. For the poor, the government is a provider of services, even if the quality of these services is sometimes below par. For the rich, the government is often an inefficient and corrupt obstacle blocking their way. For the shopkeeper, government is the tax collector demanding bribes; for the salaried employee, it is the authority that claims a large chunk of the monthly pay cheque while providing little to nothing in return.

These answers are all true, but they are also all false. That is because, much like the fable of the six blind men and the elephant, our view of government depends upon which part of it we interact with. Putting all the pieces together, we believe that the fundamental nature of government is a platform, where an entire nation comes together and designs laws and institutions meant to channel resources towards the greatest good of every citizen. In turn, every citizen has the right to question this system when it fails to deliver.

Over the years, our government's focus has evolved from building the infrastructure of a new nation, to fighting famines with the Green Revolution, to the nationalization and subsequent liberalization of the economy. This shifting spectrum of priorities holds a mirror to the needs of the people in each era, and government has had to evolve in order to meet each of these challenges effectively.

Governing with data

In the pre-technology era, the most common way for the government to achieve a particular outcome was through the creation of a public sector institution. Easy access to affordable fuel? Create public sector oil marketing companies. Food procurement and guaranteed minimum grain prices for farmers? Enter the Food Corporation of India. Bank accounts and credit for all? Nationalized banks are the answer. But is this the only model that works?

We believe it is time for government to adopt a new model—that of governing with data. How would this work in practice? Let's take the case of LPG, where the government offers a subsidy to ensure easy access to affordable fuel. Today, the entire LPG subsidy scheme has been transformed thanks to a thin layer of technology. Customers pay market price for LPG, and the subsidy amount is transferred directly to their bank accounts. It becomes easy to trace which person is availing of the subsidy, and how many cylinders they are consuming.

This data can provide any number of useful insights to the government—the average waiting time to obtain a cylinder, local consumption patterns, seasonal trends, and so on. If an area is hit by a natural disaster, all the government has to do is push a button, and afflicted residents will receive a higher subsidy as needed. Similarly, citizens with a PAN card who pay a certain amount of income tax may be able to afford the market price for LPG, and once identified by the system, they might stop receiving a subsidy altogether. No longer does the government need to own public sector institutions to achieve this goal. It can do so simply through a platform, where any business can participate, and policies are made and implemented based on data.

A simple way to achieve nationwide banking coverage is for the government to declare that all benefit transfers will be made directly into bank accounts. It can mandate that banks must open accounts for everyone, and it can ask for regular reports on the number of accounts opened, the total funds deposited and withdrawn, the average balance in an account, and so on. Based on this data, the government can give banks an incentive to operate in remote and rural areas by raising the transaction fees, while encouraging competition in more lucrative urban markets and driving the goal of universal access to financial services.

A critical requirement for governance by data is that these numbers need to be gathered in real time. Yearly statistics have no meaning in the new world. We need data flowing in from every corner of the country, updated every second. By correctly structuring incentives, leveraging the power of markets, and designing robust technology solutions generating real-time data, entire bureaucracies can be accommodated on a central dashboard. It's easy to envision a Network Operations Centre (NOC) as the nerve centre of the government, tracking, monitoring and measuring administrative processes as they happen.

The platforms that have been created in the last few years will fundamentally change the interaction between citizens, markets and the state. While individually powerful, together they deliver value greater than the sum of their parts. As a digital identity, Aadhaar allows for electronic identity authentication. e-KYC obviates the need for submitting paper documents and photos. The e-Sign platform allows people to endorse any document with a digital signature, guaranteeing the safety and authenticity of electronic documents. The Digital Locker platform is a secure repository for all digitally signed documents, from degree certificates to property documents. The National Payments Corporation of India has launched the Unified Payment Interface, allowing for the processing of mobile-based payments for any service. Taken together, these developments will empower the citizen to manage his life and all his interactions with the government digitally. In effect, government will disappear from people's everyday lives; instead of taking the physical form of offices and bureaucrats, government will now be evident only through the delivery of its services and their outcomes.

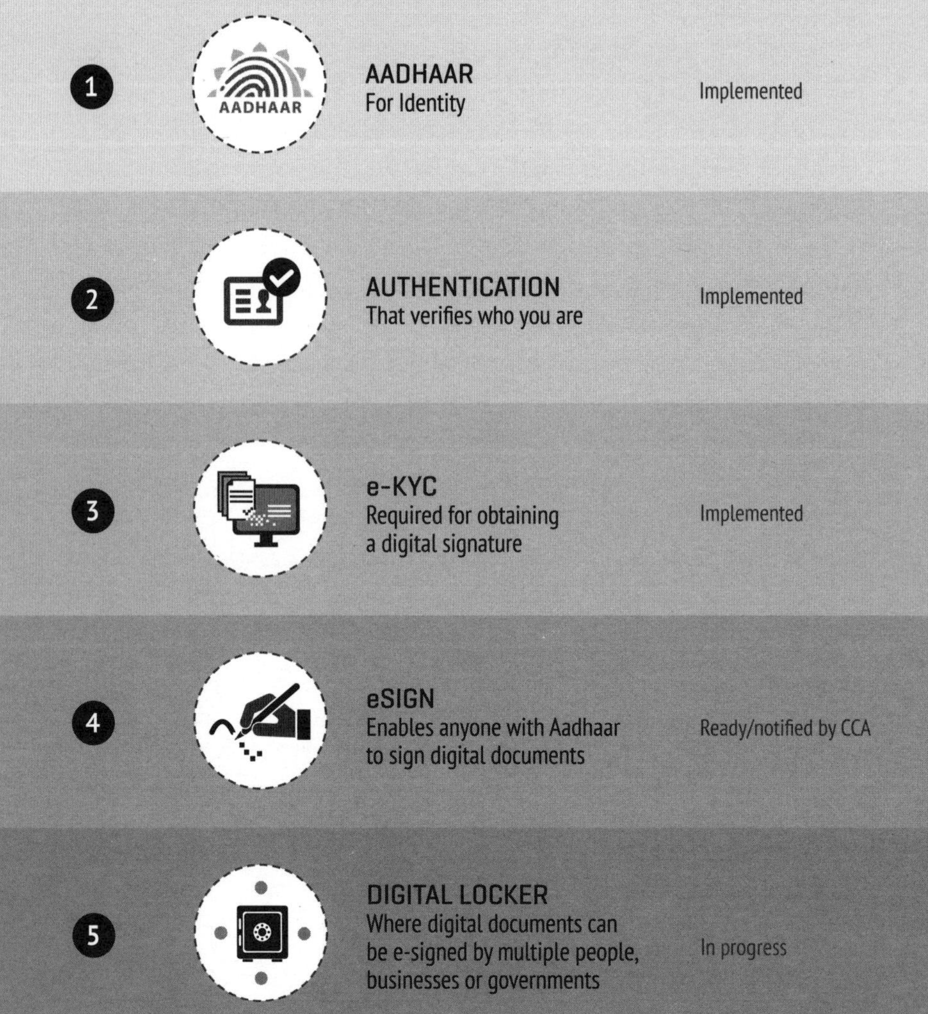
1
AADHAAR
AADHAAR
For Identity
Implemented
2
AUTHENTICATION
That verifies who you are
Implemented
3
e-KYC
Required for obtaining
a digital signature
Implemented
4
eSIGN
Enables anyone with Aadhaar
to sign digital documents
Ready/notified by CCA
5
DIGITAL LOCKER
Where digital documents can
be e-signed by multiple people,
businesses or governments
In progress

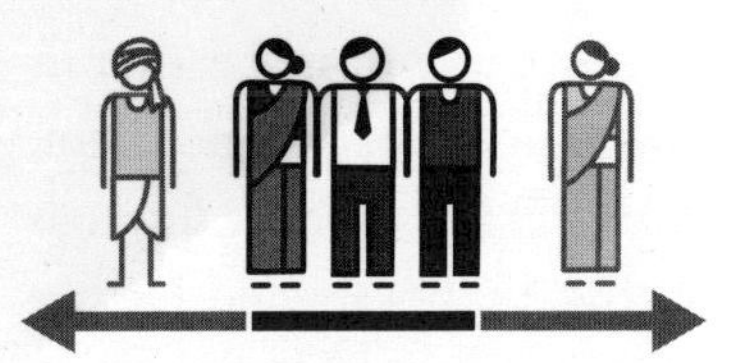

INCLUSIVE

Cost of transactions can be reduced to almost zero, widening access to government services.

SAFE STORAGE

Can be stored securely for a long period of time.

TRUST

With digital signatures, there is very little room for forgery.

SPEED

Paperwork that takes days or weeks can be concluded in a matter of minutes.

5 PLATFORMS THAT ENABLE A PAPERLESS SOCIETY

PRO-ENVIRONMENT

Use of digital documents at scale will reduce paper consumption and waste.

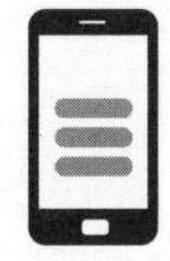

CONVENIENCE

Services on all 5 platforms will be accessible on smartphones.

Let's revisit Dharavi's Pitkekar family, who we mentioned earlier in the book, and take stock of the socio-economic climate around them. Mallava Pitkekar, who worked for 2 cents (Rs 1.25) a day sorting through trash, probably had a vision for the future that did not extend further than the holy trinity of the 1970s—'*roti, kapda aur makaan*' (food, clothing and housing). A simple calculation tells us that, not accounting for inflation, the entire GDP of India in 1970 was equivalent to just two years of social security expenditure in today's India. As Mallava's son Rayappa began the climb to prosperity with his leather business, the economic liberalization of the 1990s began. No longer were people content with the bare necessities; they wanted a better quality of life, encapsulated in the new trinity of '*bijli, sadak, pani*' (electricity, roads, water). Now, Rayappa's children's ambitions far outstrip the current pace of economic growth; we believe theirs is the generation who will espouse the digital mantra of 'Jan Dhan, Aadhaar, Mobile'—the JAM trinity— as the key to a promising future.

If Mallava received a government pension, it would be easy for the administration to track whether money had been deposited in her account, and was duly withdrawn. In Rayappa's case, the data from his business transactions captured by the GSTN could be used to assess his creditworthiness, and help to secure him a loan. The government would play a more indirect role here, regulating an ecosystem of data sharing, credit ratings and risk distribution. When it comes to Rayappa's children, if they receive education scholarships, the money can be directly transferred into their bank accounts. As they pursue advanced studies, their degree certificates, a proof of their human capital, will be available in a digital locker.

All these data platforms would make it possible for the government to provide individualized services to every citizen, rather than monitoring only aggregate outcomes; e-commerce companies already operate on this model. If we want to leapfrog over our existing problems to become a major player on the global stage, this is the only route we can take. However, we reiterate that business as usual simply will not do. One and a quarter billion people are asking for change, and they want it now; technology is the only answer to their questions. Traditionally, government has bought and used technology simply as a tool to automate processes—maintaining electronic instead of paper records, for example—but the citizen experience remains largely unaltered. This is not what we mean by a technology-enabled solution. We are talking about radically reimagining government, its purpose, its role and the way it carries out its functions, with technology at its core.

Fighting graft with systems

Every attempt at reform must sooner or later tackle the endemic corruption that bedevils our administrative machinery. Virtually every day, newspaper headlines shriek about yet another scam unearthed—2G, Saradha, Vyapam, the list goes on—which then quickly descend into a welter of accusations traded back and forth. A closer look reveals that every one of these scams can be traced to an underlying systemic issue. With a well-planned, auction-based system to allot spectrum licences to telecom companies, there would be no 2G scam. With universal financial inclusion through formal systems, there would be no Saradha scam. With enough colleges offering quality education and with enough employment opportunities for qualified individuals, there would be no Vyapam scam.

Even now, scams can be averted by creating stronger systems. It doesn't take a crystal ball to predict where scams will occur—they happen when the government is a buyer, seller or regulator capable of causing huge profits or losses to firms at the stroke of a pen. They happen when goods are sold at subsidized prices ostensibly to benefit the needy, a situation ripe for exploitation and diversion. They happen when there are shortages—a shortage of seats in schools and colleges, for example, or a shortage of jobs. They happen when ambiguous rules around taxation and compliance create opportunities for bribes. They happen when products and services become monopolies. They happen when the government, rather than the market, sets prices. Every one of these distortions, and many more besides, can be resolved with the use of data and technology.

Keeping data safe

To some people, a future in which technology assumes such a primary role is an Orwellian dystopia. How do we safeguard ourselves against such a scenario? We must first realize that it's not just the government's data collection systems we should be concerned about; private organizations also collect and analyse large amounts of consumer information. While government must operate under the strict set of rules to which it is bound, it is essential that a strong privacy and data protection law be put in place. Some also fear that the government can use this data against its own people. Given that the government holds the monopoly on creating laws, this is a valid concern.

Here is where we must turn our attention towards the design of our institutions, the separation of powers, and the checks and balances that prevent the system from running amok. Data protection laws should defend not only the individual's right to privacy but also confer protection from the government itself. However, we must ultimately recognize that one cannot operate within government to bring about widespread change with a mindset of complete mistrust. Healthy scepticism aside, we must put our faith in a system that has worked for the last sixty-eight years, and shift our focus to creating institutions that deliver.

An app economy

India is often said to be beset by the 'curse of informality'. Nearly three-fourths of all employment is in the informal sector[3]—millions of farmers tilling tiny patches of land, retailers running cramped shops, women selling vegetables by the roadside. A common solution to 'formalize' our economy is to adopt the industrial model, building organized factories staffed by thousands of employees, creating a network of organized retail outlets, and promoting large organized farms, which can run into thousands of hectares. This is the route the western world took in the twentieth century. But is this really a twenty-first century plan for India?

With the advent of technology, we think there is an entirely new way of formalizing the economy. With smartphones becoming increasingly ubiquitous, we can harness the power of cloud computing, big data and analytics to move towards a different kind of aggregation, one that is gaining traction in the private sector. Today, taxi services like Ola and Uber are aggregators, organizing thousands of individual drivers on a single platform. By pooling the homes and spare bedrooms of thousands of people, Airbnb now has more rooms than the biggest hotel chains. In India, Oyo Rooms has achieved much the same with budget hotels. Flipkart and Amazon provide marketplaces where merchants sell just about anything to hundreds of millions of customers.

In his pioneering article, 'The Nature of the Firm', written in 1937, the economist Ronald Coase argued that the costs of carrying out transactions—the costs of search and information, coordination and contracting—meant that it made better financial sense for people to organize themselves into firms. As the friction around these costs grew, firms themselves would keep expanding. While this was an accurate worldview in 1937, today technology has upended Coase's law. Perhaps India can retain the small retailer, the small farmer and the small entrepreneur; instead of converting them into faceless employees of large firms, we can bring them 'on the grid'. With their smartphones, they will have access to the best technology, supplies and information, bringing markets closer and allowing them to make a living for themselves.

All the government needs to do is to create a strong regulatory regime, and operate effective social safety nets. A million farmers, a million entrepreneurs and a million retailers, organized through technology into virtual platforms, would be a very potent force promoting job creation, innovation and creativity. This approach ties in neatly with our constraints as a nation, whether it's the scarcity of land in urban areas, the difficulty of land acquisition and aggregation, or simply the context of our political economy and bottom-up democracy. If we are to usher India into a new era of growth, the way we organize our people and their economic principles needs to be reimagined from the ground up.

A new federal structure

Not just our economy but the very structure of our government needs a rethink. Gone are the days when the centre used money as a stick to get the states in line. While the Constitution of India recognizes states as equal partners, the balance of power tilted away from them in the past, making them increasingly dependent on the centre for funds. Today, many of our states are well run, and with the formation of the NITI Aayog and the upcoming GST reform, they will achieve further fiscal independence. Money is no longer the motive force for change—you can't drive reform through fiscal initiatives. Under the latest budget and the recommendations of the fourteenth Finance Commission, the centre will share 42 per cent of its revenue with the states, as opposed to 32 per cent in 2014–15.[4] So, a significantly larger chunk of funds will now be provided by the centre to the states as untied funds.

We recollect an example from a state revenue officer. He explained that in a particular year, the amount allocated to the state's midday-meal scheme by the centre was only 80 per cent of the total budget. What was the state to do now? Could it cite the budget shortfall as a reason to stop serving meals to students? Obviously not. One solution would be to channel funds from another budget head to the midday-meal scheme to keep it going. But money received from the centre is usually earmarked for a specific purpose, and cannot be used elsewhere, no matter how pressing the need. These issues of fund flow can be resolved by cutting down the number of schemes and running them more efficiently; equally important, a greater proportion of central funds should be untied, eliminating restrictions on how this money can be spent and allowing states to manage their money more effectively.

Increasingly, India is getting to a point where money itself is no longer the bottleneck. The finances of many states are quite robust. The centre–state relationship has evolved to the point where we believe that the centre must provide value beyond money through world-class platform development. States should be free to opt for these common platforms because they see a clear benefit in participation, rather than through the carrot of money or the stick of legislation. The GSTN is one such partnership. Aadhaar and the redesign of social security schemes is another such partnership. Electronic toll, once it goes beyond national highways to become a common platform for all transport-related payments, will become another key initiative. The Expenditure Information Network, once operational, will streamline fund flow, utilization and monitoring. Every platform we have discussed in this book has the potential to reshape the centre–state–local government partnerships and give us a new form of governance. This, we believe, is what true cooperative federalism is all about.

Rebooting India

For most people, interacting with the state is a stressful and arduous task. Standing in long queues, the lack of access to information, waiting endlessly for rations, negotiating with corrupt officials—these are just some of the irritants, both major and minor, that people must deal with. Is there a way we can smoothen out these bumps, in effect rendering the state 'invisible' to the people? The ideas in this book lay out a concrete road map to achieve this vision.

The biggest barrier to the ideas we propose in this book is mindsets. In a system that clings tenaciously to hierarchy, it is hard to recognize that a twenty-five-year-old 'techie' might have better ideas than the veteran official in his fifties. That value and knowledge lie not at the top of a silo, but at the boundaries across varied disciplines. That problem-solving is not about big budgets and a cast of thousands, but small teams with shoestring budgets—teams that include technologists, social activists, people who have built successful businesses, domain experts and bureaucrats. That we have to try often and fail fast without getting bogged down by the armchair experts and critics ready to pounce on the first sign of failure. That it is easier to find fraud by analysing data rather than harassing citizens. That corruption can be eliminated not by having more laws, more inspectors and more ombudsmen, but by reimagining the processes that allow corruption to flourish in the first place. That it is better to have no entry barriers

10 GRAND CHALLENGES

6

Long-term, aspirational

Elections

Education

Healthcare

Energy

Justice

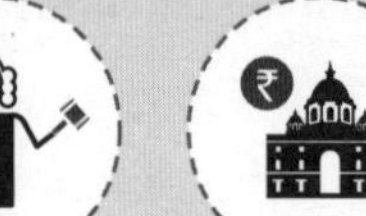

Expenditure

4

Early stages, to be scaled up

Cashless/ MicroATM

Paperless society/ e-KYC

Electronic Toll

Goods and Services Tax

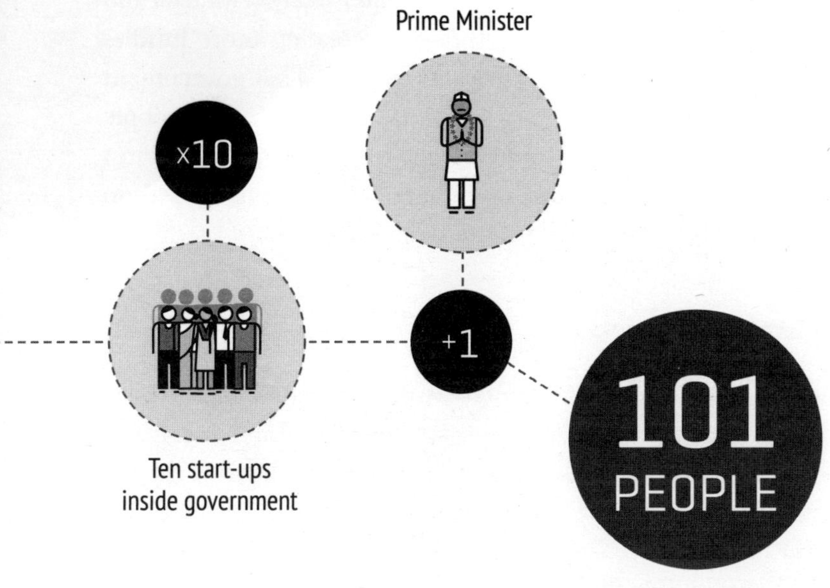
Prime Minister
×10
+1
Ten start-ups
inside government
101
PEOPLE
Transform India and help
realize a billion aspirations

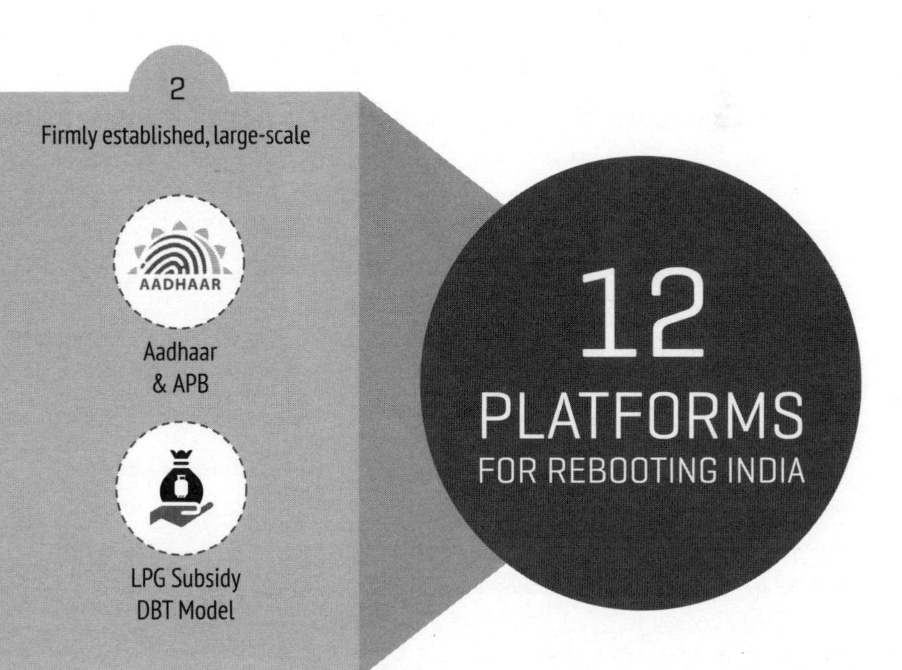
2
Firmly established, large-scale
AADHAAR
Aadhaar
& APB
LPG Subsidy
DBT Model
12
PLATFORMS
FOR REBOOTING INDIA

to government systems, allowing people to enter freely with data and technology used to detect fraud, rather than creating more hurdles for people wanting access to benefits and services. That government should build minimalistic platforms and let the creativity of the people drive innovation. That the best ideas can be found anywhere, not just in the halls of government. That the best solutions are not based on precedents from the past, but are rooted in anticipation of the future.

Big problems, small teams

The success of any project, whether in government or elsewhere, depends on two things: the right people for the job, and an institutional framework that provides the right incentives for everyone in the ecosystem. We firmly believe, based on our experience with Aadhaar, that all we need is a start-up in government to address every grand challenge we face. Equally important, the start-up must eventually integrate into the main body of the government rather than continuing to remain a separate unit; this is the only way to ensure that these projects survive and thrive in government while minimizing the potential for friction.

In order to ensure that these projects do not sink under the weight of bureaucratic gravity, it is essential that they be anchored under the national leader, the prime minister. When we look at some of India's earlier government start-ups, what stands out is the close rapport between the prime minister and the leader of the initiative—the warm relationship between Jawaharlal Nehru and Dr Homi Bhabha, who led India's atomic energy programme, is one example. It is this kind of dedicated support at the highest levels that allows for entrepreneurial projects within government to succeed.

In markets, competition drives change. A new competitor can offer a better product, and replace an old monopoly. In politics, change is driven through the electoral process. In order to be successful, politicians have to be able to deliver on the people's aspirations. As the world evolves with technology, government has to keep pace, and it is up to the political masters to make this happen.

We have identified twelve grand challenges in this book; the first two of these—Aadhaar and PaHal—have already been scaled successfully. That leaves us with ten more grand challenges for which we need ten start-ups in government, each with a team of ten dedicated multidisciplinary champions. Such teams, operating under the authority of the prime minister, can drive the sweeping transformation and innovative thinking capable of fulfilling a billion aspirations.

We are much better off dreaming, taking risks, and trying to realize a billion aspirations; at best we risk falling flat on our faces. Far more egregious, and most dangerous to our country, is going about 'business as usual', leaving a billion voices unheard and a billion frustrations unresolved.

NOTES

Introduction

1. Ghoshal, Devjyot. 11 April 2012. 'How Aadhar is transforming lives across Jharkhand'. Rediff.com.
http://www.rediff.com/money/slide-show/slide-show-1-column-aadhar-is-transforming-lives-across-jharkhand/20120411.htm
2. Golden Quadrilateral Highway Network, India.
http://www.roadtraffic-technology.com/projects/golden-quadrilateral-highway-network/
3. Bagla, Pallava. 23 September 2014. 'Mangalyaan, the cheapest Mars mission ever'. NDTV.
http://www.ndtv.com/india-news/mangalyaan-the-cheapest-mars-mission-ever-66974
4. Rashtriya Swasthya Bima Yojana.
http://www.rsby.gov.in/about_rsby.aspx
5. Andreessen, Marc. 20 August 2011. 'Why Software Is Eating the World'. *The Wall Street Journal*.
http://www.wsj.com/articles/SB10001424053111903480904576512250915629460
6. Saran, Rohit. 26 December 2005. '1995: Cell phones arrive'. *India Today*.
http://indiatoday.intoday.in/story/bengal-cm-jyoti-basu-made-indias-first-cell-phone-call-to-telecom-minister-sukh-ram-in-1995/1/192421.html
7. Telecommunications statistics in India:
https://en.wikipedia.org/wiki/Telecommunications_statistics_in_India
Telecom Regulatory Authority of India. 4 July 2012. 'Highlights on Telecom subscription data as on 31st May 2012'.

http://www.trai.gov.in/WriteReadData/PressRealease/Document/PR-TSD-May12.pdf

Press Trust of India. 29 January 2014. 'India to have 243 million Internet users by June 2014: Report'. NDTV.

http://www.ndtv.com/india-news/india-to-have-243-million-internet-users-by-june-2014-report-549211

8. 'Life in a slum'. BBC News.
   http://news.bbc.co.uk/2/shared/spl/hi/world/06/dharavi_slum/html/dharavi_slum_into.stm
9. Varghese, Gigil, Zachariah, Mini Pant, and Dagli, Kinjal. 17 January 2009. 'Slum Gods'. *Hindustan Times.*
   http://www.hindustantimes.com/india/slum-gods/story-QBMRbH33hPfZnBkVlVo81I.html
10. Press Trust of India. 15 March 2012. 'Economic Survey 2011–12: India to join league of youngest nations by 2020'. *Economic Times.*
    http://articles.economictimes.indiatimes.com/2012-03-15/news/31197250_1_demographic-dividend-economic-survey-life-expectancy
11. World Bank Group. Ease of doing business rankings.
    http://www.doingbusiness.org/rankings
12. Micklethwait, J., and Wooldridge, A. 2014. *The Fourth Revolution: The Global Race to Reinvent the State.* Penguin Books.
13. Guha, Ramachandra. 2007. *India After Gandhi: The History of the World's Largest Democracy*. Pan Macmillan.
14. Anand, Umesh. September 2012. 'Ratnauli's wired hero'.
    http://www.civilsocietyonline.com/pages/HOF_Details.aspx?185
15. Growth of Facebook.
    http://en.wikipedia.org/wiki/Template:Facebook_growth
16. Shah, Ajay. 2006. 'Improving Governance Using Large IT systems'. *Documenting Reforms: Case Studies from India.* S. Narayan (ed.). New Delhi: Macmillan India, pp. 122–148.
    NSDL charges.
    https://nsdl.co.in/about/charges.php
17. 2001 Census of India: Migration.
    http://censusindia.gov.in/Census_And_You/migrations.aspx
18. Shah, Amrita. 2007. *Vikram Sarabhai: A Life.* New Delhi: Penguin Books.

19. Transcript of Federalist Papers, No. 10 & No. 51 (1787–1788). http://www.ourdocuments.gov/doc.php?flash=true&doc=10&page=transcript
20. Pritchett, Lant. 2009. 'Is India a Flailing State?: Detours on the Four Lane Highway to Modernization'. HKS Faculty Research Working Paper Series, RWP09-013, John F. Kennedy School of Government, Harvard University.

1. Aadhaar: From zero to a billion in five years

1. 30 September 2010. 'Ranjana Sonawane is now a 12-digit no.'. *Times of India.*
   http://timesofindia.indiatimes.com/india/Ranjana-Sonawane-is-now-a-12-digit-no-/articleshow/6654987.cms
   Rabade, Parag. 2 October 2010. 'Ranjana, the UID of Aadhaar'. *Deccan Herald.*
   http://www.deccanherald.com/content/101472/ranjana-uid-aadhaar.html
   29 September 2010. 'Meet the nation's first UID holders'. Rediff.com.
   http://www.rediff.com/news/report/no-one-will-cheat-me-now-uid-recepient-tembhali/20100929.htm
   Pal, Chandrima. 30 September 2010. 'For Tembhli, power, paint and an identity'. *Mumbai Mirror.*
   http://www.mumbaimirror.com/mumbai/cover-story/For-tembhli-power-paint-and-an-identity/articleshow/16057599.cms
2. Unique Identification Authority of India, Public Data Portal. http://portal.uidai.gov.in
3. Parker, Ian. 3 October 2011. 'The I.D. Man'. *New Yorker.* http://www.newyorker.com/reporting/2011/10/03/111003fa_fact_parker.
4. UNICEF India Statistics. http://www.unicef.org/infobycountry/india_statistics.html.
5. April 2010. 'UIDAI Strategy Overview: Creating a Unique Identity Number for Every resident in India'. Unique Identification Authority of India (UIDAI), Planning Commission, Government of India.
6. Nilekani, Nandan. 2008. *Imagining India: Ideas for the New Century.* New Delhi: Penguin Books.

7. Nilekani, Nandan. January–March 2014. 'The Untold Story of Aadhaar'. Inclusion.
http://inclusion.skoch.in/story/10/the-untold-saga-of- aadhaar-310.html
8. PTI. 12 January 2012. 'Planning Commission, Home Ministry in tussle over UID project'. *Economic Times*.
http://articles.economictimes.indiatimes.com/2012-01-20/news/30647230_1_uidai-home-ministry-aadhaar-numbers
9. Unique Identification Authority of India, Public Data Portal Dashboard for enrolments.
https://portal.uidai.gov.in/uidwebportal/dashboard.do.
10. Mehdudia, Sujay. 18 February 2014. 'Government fully committed to Aadhaar'. *The Hindu*.
http://www.thehindu.com/news/national/government-fully-%20committed-to-aadhaar/article5699113.ece
11. 29 August 2011. 'Plan panel, parliamentary committee red flag rising UID costs'. *Indian Express*.
http://archive.indianexpress.com/news/plan-panel-parliamentary-committee-red-flag-rising-uid-costs/838507/0
12. UIDAI Working Papers. 'Ensuring uniqueness: Collecting iris biometrics for the unique ID mission'.
https://uidai.gov.in/UID_PDF/Working_Papers/UID_and_iris_paper_final.pdf
13. Anand, Utkarsh. 25 March 2014. 'Don't exclude those without Aadhaar, share data: SC'. *Indian Express*.
http://indianexpress.com/article/india/india-others/supreme-court-bars-sharing-of-uidai-info-denial-of-benefits-to-those-without-aadhaar/
14. Standing Committee on Finance (2011–12), fifteenth Lok Sabha, Ministry of Planning. December 2011. 'The National Identification Authority of India Bill, 2010'. Forty-Second report.
http://164.100.47.134/lsscommittee/Finance/42%20Report.pdf
15. 11 September 2014. 'Government approves fifth phase of Aadhaar enrollment in UP, Bihar, Jharkhand and Chhattisgarh'. *Economic Times*.
http://articles.economictimes.indiatimes.com/2014-09-11/news/53811399_1_uidai-mantri-jan-dhan-yojana-aadhaar-enrollment
16. Nair, Pravin. 20 April 2014. 'First Aadhaar card owner struggles for a living'. *Hindustan Times*.

http://www.hindustantimes.com/india/first-aadhaar-card-owner-struggles-for-a-living/story-4XDs3Ffu0aoG8NCPRWDj3O.html

2. Aadhaar: Behind the scenes

1. Parker, Ian. 3 October 2011. 'The I.D. Man'. *New Yorker.*
http://www.newyorker.com/magazine/2011/10/03/the-i-d-man
2. Thimmayya, Daniel. 5 April 2015. 'Trafficked Women, Orphans Finally Get an Aadhar of Existence in Tamil Nadu'. *New Indian Express.*
http://www.newindianexpress.com/thesundaystandard/Trafficked-Women-Orphans-Finally-Get-an-Aadhar-of-Existence-in-Tamil-Nadu/2015/04/05/article2747850.ece
3. Kumar, Raj. 16 December 2012. 'See how Aadhaar benefits women'. *Telegraph.*
http://www.telegraphindia.com/1121216/jsp/jharkhand/story_16321510.jsp#.VZ51JxOqqko
4. December 2009. 'Biometric Design Standards for UID Applications (Version 1.0)'. Unique Identification Authority of India.
http://uidai.gov.in/UID_PDF/Committees/Biometrics_Standards_Committee_report.pdf
5. Krebs, Brian. 6 July 2009. 'Researchers: Social Security Numbers Can Be Guessed'. *Washington Post.*
http://www.washingtonpost.com/wp-dyn/content/article/2009/07/06/AR2009070602955.html
6. Radicati, Sara. April 2013. 'Email Statistics Report, 2013–2017'. The Radicati Group.
http://www.radicati.com/wp/wp-content/uploads/2013/04/Email-Statistics-Report-2013-2017-Executive-Summary.pdf
7. 'Google Search Statistics'. Internet Live Stats.
http://www.internetlivestats.com/google-search-statistics/
8. 'Number of monthly active Facebook users worldwide as of 2nd quarter 2015 (in millions)'. Statista.
http://www.statista.com/statistics/264810/number-of-monthly-active-facebook-users-worldwide/
9. Anders, George. 19 February 2014. 'Facebook's $19 Billion Craving, Explained By Mark Zuckerberg'. *Forbes.*

http://www.forbes.com/sites/georgeanders/2014/02/19/facebook-justifies-19-billion-by-awe-at-whatsapp-growth/

3. Banking on government payments

1. 2013. 'Report of the Committee on Comprehensive Financial Services for Small Businesses and Low Income Households'. Reserve Bank of India.
https://rbidocs.rbi.org.in/rdocs/PublicationReport/Pdfs/CFS070114RFL.pdf
2. Ghosh, Palash. 6 June 2013. 'How many people pay income tax? Hardly anyone'. *International Business Times*.
http://www.ibtimes.com/how-many-people-india-pay-income-tax-hardly-anyone-1294887
3. Yardley, Jim. 28 December 2011. 'In One Slum, Misery, Work, Politics and Hope'. *New York Times*.
http://www.nytimes.com/2011/12/29/world/asia/in-indian-slum-misery-work-politics-and-hope.html
4. Burman, Abheek. 25 October 2013. 'Informal workers, making up 90% of workforce, won't get a good deal till netas notice them'. *Economic Times*.
http://articles.economictimes.indiatimes.com/2013-10-25/news/43395491_1_neelkanth-mishra-india-fall-informal-economy
5. 28 September 2013. 'India's Informal Economy. Hidden Value'. *Economist*.
http://www.economist.com/news/asia/21586891-activities-out-sticks-may-add-more-gdp-was-thought-hidden-value
6. Dr G. Rajan, Raghuram, Governor, Reserve Bank of India. 11 December 2013. Talk at the Delhi Economics Conclave titled 'Financial Sector Reforms'.
http://www.rbi.org.in/scripts/BS_SpeechesView.aspx?Id=863
7. Collins, Daryl, Morduch, Jonathan, Rutherford, Stuart, Ruthven, Orlanda. 2009. *Portfolios of the Poor: How the World's Poor Live on $2 a Day*. Princeton: Princeton University Press.
8. February 2012. 'Report of the Task Force on an Aadhaar-Enabled Unified Payment Infrastructure'. Ministry of Finance, Government of India.

http://finmin.nic.in/reports/Report_Task_Force_Aadhaar_PaymentInfra.pdf

9. April 2010. 'From Exclusion to Inclusion with 'Micropayments'. Unique Identification Authority of India.
http://uidai.gov.in/UID_PDF/Front_Page_Articles/Strategy/Exclusion_to_Inclusion_with_Micropayments.pdf
10. October 2010. 'Growth, financial inclusion, and a $22.4 billion savings'. McKinsey & Company.
http://mckinseyonsociety.com/inclusive-growth-and-financial-security/
11. Wolman, David. 2012. *The End of Money: Counterfeiters, Preachers, Techies, Dreamers—and the Coming Cashless Society*. Boston: Da Capo Press, p. 353.
12. 2015. 'Pradhan Mantri Jan Dhan Yojana'. Department of Financial Services, Ministry of Finance, Government of India.
http://pmjdy.gov.in/account-statistics-country.aspx
13. Thornycroft, Peta. June 2006. 'The $100,000 note that won't buy a loaf of bread in Zimbabwe'. *Telegraph*.
http://www.telegraph.co.uk/news/worldnews/africaandindianocean/zimbabwe/1520590/The-100000-note-that-wont-buy-a-loaf-of-bread-in-Zimbabwe.html
14. Gustafsson, Katarina, and Magnusson, Niklas. 28 October 2013. 'Stockholm's Homeless Accept Cards as Cash No Longer King'. *Bloomberg*.
http://www.bloomberg.com/news/2013-10-27/stockholm-s-homeless-accept-cards-as-cash-no-longer-king.html
15. 14 November 2013. 'Mobile Money: The Opportunity for India'. Mobile Money Association of India (MMAI) and GSMA submission to the Reserve Bank of India.
16. Ghosh, Suprotip. 8 December 2013. 'Banking the Unbanked'. *Business Today*.
http://businesstoday.intoday.in/story/rbi-expansion-general-public-helps-take-banking-rural-india/1/200646.html
17. February 2015. 'Unified Payments Interface. API and Technology Specifications (Version 1.0)'. National Payments Corporation of India.
http://www.npci.org.in/documents/API_Technology_Specifications_Version_1.pdf

4. Mending our social safety nets

1. 12 February 2012. UP farmer on fertilizer availability.
   https://www.youtube.com/watch?v=YBATYKn_cBY
2. 10 June 2008. 'Karnataka farmers on rampage over fertilizer shortage'. *Times of India.*
   http://timesofindia.indiatimes.com/india/Karnataka-farmers-on- rampage-over-fertilizer-shortage/articleshow/3115597.cms
3. Green Revolution. Wikipedia.
   http://en.wikipedia.org/wiki/Green_Revolution
4. Mukherji, Biman. 8 July 2014. 'India to Keep Grain Supplies on Fears of Poor Monsoon Rains'. *Wall Street Journal.*
   http://www.wsj.com/articles/india-to-keep-grain-supplies-on-fears-of-poor-monsoon-rains-1404814017
   Statistics on Agricultural Production, FAOSTAT.
   http://faostat.fao.org/site/339/default.aspx
5. Ministry of Law and Justice. The National Food Security Act, 2013.
   http://indiacode.nic.in/acts-in-pdf/202013.pdf
6. 'Wiping every tear from every eye': the JAM Number Trinity Solution.
   http://indiabudget.nic.in/es2014-15/echapvol1-03.pdf
7. Bajaj, Vikas. 7 June 2012. 'As Grain Piles Up, India's Poor Still Go Hungry'. *New York Times.*
   http://www.nytimes.com/2012/06/08/business/global/a-failed-food-system-in-india-prompts-an-intense-review.html
8. 17 March 2012. 'Let them eat baklava'. *Economist.*
   http://www.economist.com/node/21550328
   Friedman, Thomas L. 7 April 2012. 'The Other Arab Spring'. *New York Times*, Sunday Review Op-Ed.
   http://www.nytimes.com/2012/04/08/opinion/sunday/friedman-the-other-arab-spring.html?_r=0
9. Myers, Norman. 9 August 2007. 'Perverse subsidies'. *The Encyclopedia of Earth.*
   http://www.eoearth.org/view/article/155197/
10. World Bank Data, fertilizer consumption (kilograms per hectare of arable land).
   http://data.worldbank.org/indicator/AG.CON.FERT.ZS

11. Anand, Geeta. 22 February 2010. 'Green Revolution in India Wilts as Subsidies Backfire'. *Wall Street Journal.* http://www.wsj.com/news/articles/SB10001424052748703615904575052921612723844
12. Padma, T.V. 13 August 2009. 'Thirsty Indian farming depleting water resources'. SciDevNet. http://www.scidev.net/global/farming/news/thirsty-indian-farming-depleting-water-resources.html
13. Lahariya, Khabar and Tewari, Saumya. 10 July 2015. 'Mathuranpura And The World's Worst Groundwater Crisis'. IndiaSpend. http://www.indiaspend.com/cover-story/mathuranpura-and-the-worlds-worst-groundwater-crisis-87432
14. Zhong, Raymond. 17 July 2014. 'This Is Why India Has to Shrink the Subsidy Raj'. http://blogs.wsj.com/indiarealtime/2014/07/07/this-is-why-india-has-to-shrink-the-subsidy-raj/
15. Poovanna, Sharan, and Saha, Samiran. 19 May 2013. 'Car makers run into diesel dilemma'. http://www.newindianexpress.com/business/news/Car-makers-run-into-diesel-dilemma/2013/05/19/article1596282.ece

    Davies, Alex. 22 August 2013. '10 Reasons You Should Buy A Diesel Car'. *Business Insider India.* http://www.businessinsider.in/10-Reasons-You-Should-Buy-A-Diesel-Car/articleshow/21967160.cms
16. Gelb, Alan and Diofasi, Anna. 17 March 2015. 'Ghostbusters: Linking Subsidy Reform and Biometric Identification in India'. Center for Global Development. http://www.cgdev.org/blog/ghostbusters-linking-subsidy-reform-and-biometric-identification-india#_ftn1
17. Dehejia, Rupa Subramanya. 20 June 2011. 'Economics Journal: How the Oil Subsidy Fuels Corruption and Death'. *Wall Street Journal.* http://blogs.wsj.com/indiarealtime/2011/06/20/economics-journal-how-the-oil-subsidy-fuels-corruption-and-death/
18. Narayan, Hari. 5 January 2013. 'The extraordinary tale of an ordinary man'. *The Hindu.* http://www.thehindu.com/features/magazine/the-extraordinary-tale-of-an-ordinary-man/article4269047.ece

19. 9 January 2014. 'Fuelling controversy'. *Economist.*
http://www.economist.com/news/finance-and-economics/21593484-economic-case-scrapping-fossil-fuel-subsidies-getting-stronger-fuelling
20. 12 June 2014. 'Scrap them'. *Economist.*
http://www.economist.com/news/leaders/21604170-there-are-moves-around-world-get-rid-energy-subsidies-heres-best-way-going
21. 24–25 September. G20 Leaders' Statement: The Pittsburgh Summit.
http://www.g20.utoronto.ca/2009/2009communique0925.html
22. World Bank data, military expenditure (% of GDP).
http://data.worldbank.org/indicator/MS.MIL.XPND.GD.ZS
23. June 2011. 'Interim report of the task force on direct transfer of subsidies on kerosene, LPG and fertilizer'. Ministry of Finance, Government of India.
http://finmin.nic.in/reports/Interim_report_Task_Force_DTS.pdf
24. 'Wiping every tear from every eye': the JAM Number Trinity Solution.
http://indiabudget.nic.in/es2014-15/echapvol1-03.pdf
25. Shah, Viral. 'Transparency in the LPG subsidy'. Prof. Ajay Shah's blog. (Professor, National Institute of Public Finance and Policy).
http://ajayshahblog.blogspot.in/2012/07/transparency-in-lpg-subsidy.html
26. Aji, Sowmya. 26 March 2012. 'Stink of a gas scam'. *India Today.*
http://indiatoday.intoday.in/story/karnataka-24-lakh-fake-domestic-gas-connections-uncovered/1/178073.html.
27. August 2012. 'Fossil-Fuel Subsidy Reform in India: Cash transfers for PDS, kerosene and domestic LPG'. The Energy and Resources Institute, TERI University.
http://www.iisd.org/gsi/sites/default/files/ffs_india_teri_rev.pdf
28. Press Trust of India. 1 June 2013. 'LPG subsidy directly in bank accounts from today in 18 districts'. NDTV.
http://www.ndtv.com/india-news/lpg-subsidy-directly-in-bank-accounts-from-today-in-18-districts-523967
29. Mukherjee, Promit. 3 July 2015. 'DBTL helps govt save Rs 10,000 crore as illegal LPG consumption falls'. *Mint.*
http://www.livemint.com/Industry/PGCreyRo9L9rCt3Vx3xyBO/ %20 DBTL-helps-govt-save-Rs10000-crore-as-illegal-LPG-consumpti.html
30. May 2014. 'Review of the Direct Benefit Transfer for LPG Scheme, Committee Report'. Ministry of Petroleum and Natural Gas, Government of India.

http://petroleum.nic.in/docs/dhande.pdf

31. Indo-Asian News Service. 5 February 2015. 'PAHAL Yojana will end black marketing in LPG, says PM Narendra Modi'. Firstpost. http://www.firstpost.com/business/pahal-yojana-will-end-black-marketing-in-lpg-says-pm-narendra-modi-2081573.html
32. Press Trust of India. 10 July 2014. 'Budget 2014: FM Arun Jaitley raises subsidy bill but promises overhaul of grants'. *Economic Times*. http://articles.economictimes.indiatimes.com/2014-07-10/news/51300717_1_subsidy-bill-fertiliser-subsidy-lakh-crore
33. Dr Shenoy, Bhamy V. March 2010. 'Lessons Learned from Attempts to Reform India's Kerosene Subsidy'. The International Institute for Sustainable Development. http://www.iisd.org/pdf/2010/lessons_india_kerosene_subsidy.pdf
34. Kurmanath, K.V. 6 December 2012. 'Aadhaar, PDS database link to help AP plug loopholes. *The Hindu Business Line*. http://www.thehindubusinessline.com/info-tech/aadhaar-pds-database-link-to-help-ap-plug-loopholes/article4171521.ece
35. Agarwal, Surabhi. 17 February 2015. 'Aadhaar works well where put in use'. *Business Standard*. http://www.business-standard.com/article/economy-policy/aadhaar-works-well-where-put-in-use-115021700034_1.html

    Mishra, Asit Ranjan. 3 July 2015. 'NDA govt kicks off PDS reforms with direct cash transfers'. *Mint*. http://www.livemint.com/Politics/BfeNi5AreTn1cJ8ROIzxXM/NDA-kicks-off-PDS-reforms.html
36. Press Trust of India. 14 February 2010. 'Slogan of "bijli, sadak, pani is passe": Nilekani'. *The Hindu*. http://www.thehindu.com/news/national/slogan-of-bijli-sadak-pani-is-passe-nilekani/article106404.ece
37. Press Trust of India. 13 March 2015. 'Government expenditure on Aadhaar project is Rs 5,630 crore'. *Economic Times*. http://articles.economictimes.indiatimes.com/2015-03-13/news/60086261_1_aadhaar-project-aadhaar-numbers-uidai-project
38. Sethi, Nitin, and Agarwal, Surabhi. 24 February 2015. 'Centre takes steps to convert PDS to cash transfers'. *Business Standard*. http://www.business-standard.com/article/economy-policy/centre-takes-steps-to-convert-pds-to-cash-transfers-115022400033_1.html

5. Going completely paperless with e-KYC

1. 26 February 2014. 'Axis Bank introduces e-KYC, MicroATM facility'. *The Hindu Business Line.*
   http://www.thehindubusinessline.com/companies/axis-bank-introduces-ekyc-microatm-facility/article5729216.ece
2. 20 August 2014. 'Digital India—A programme to transform India into digital empowered society and knowledge economy'. Press Information Bureau, Government of India.
   http://pib.nic.in/newsite/erelease.aspx?relid=108926
3. Rai, Suyash, Sharma, Smriti, and Sapatnekar, Sanhita. 14 September 2014. 'A dramatic cost reduction for KYC using the e-KYC API of UIDAI'. Prof. Ajay Shah's blog. (Professor, National Institute of Public Finance and Policy).
   http://ajayshahblog.blogspot.in/2014/09/a-dramatic-cost-reduction-for-kyc-using.html
4. November 2012. 'Aadhaar e-KYC Service'. Unique Identification Authority of India.
   http://uidai.gov.in/images/commdoc/ekyc_policy_note_18122012.pdf
5. 28 August 2014. 'PM Narendra Modi launches Jan Dhan Yojana; to focus on combating financial untouchability'. *Indian Express.*
   http://indianexpress.com/article/india/india-others/pm-narendra-modi-launches-jan-dhan-yojana-to-focus-on-ending-finacial-untouchability/

6. Integrating our economy with the Goods and Services Tax

1. Lakshmi, Rama. 26 December 2014. 'Is India's country-wide tax a "game-changer" for business?'. *Washington Post.*
   http://www.washingtonpost.com/world/asia_pacific/is-indias-country-wide-tax-a-game-changer-for-business/2014/12/24/b0a5eef4-4654-40d3-9297-ad6ced6451b8_story.html
2. 18 September 2015. 'India's dream of borderless trade grinds to a halt at checkpoints'. IBN Live.
   http://www.ibnlive.com/news/india/indias-dream-of-borderless-trade-grinds-to-a-halt-at-checkpoints-1102758.html
3. Sinha, Varun. 22 November 2014. 'How GST Will Change the Face of Indian Economy'. NDTV.

http://profit.ndtv.com/news/economy/article-how-gst-will-change-the-face-of-indian-economy-701413

4. October 2014. 'India Development Update: Continue Domestic Reforms and Encourage Investments'. World Bank.
http://www.worldbank.org/en/country/india/publication/development-update-domestic-reforms-encourage-investments
5. 16 January 2014. 'Narendra Modi calls for reforms amid "tax terrorism"'. *Times of India*.
http://timesofindia.indiatimes.com/city/ahmedabad/Narendra-Modi-calls-for-reforms-amid-tax-terrorism/articleshow/28866113.cms
6. 14 July 2014. 'Finance ministry wants taxman not to harass taxpayers'. *Times of India*.
http://timesofindia.indiatimes.com/business/india-business/Finance-ministry-wants-taxman-not-to-harass-taxpayers/articleshow/38346860.cms
7. Mahalingam, T.V. 6 July 2014. 'Budget 2014: What India's five manufacturing hubs expect from Modi government to kick-start growth'. *Economic Times*.
http://articles.economictimes.indiatimes.com/2014-07-06/news/51107986_1_manufacturing-sector-world-manufacturing-hubs
8. 'The World Bank Indicators'.
http://data.worldbank.org/indicator/GC.TAX.TOTL.GD.ZS.
9. July 2004. 'Report of the Task Force on Implementation of the Fiscal Responsibility and Budget Management Act, 2003'. Ministry of Finance, Government of India.
http://finmin.nic.in/reports/FRBM_report.pdf
15 December 2009. 'Report of the Task Force on Goods and Services Tax'. Thirteenth Finance Commission, Government of India.
http://rajtax.gov.in/vatweb/download/gst/13th%20FCR.pdf
10. Chidambaram, P. 14 June 2015. 'The GST Bill is history in the making'. *Financial Express*.
http://www.financialexpress.com/article/fe-columnist/the-gst-bill-is-history-in-the-making-p-chidambaram/84453/
11. 31 January 2011. 'Report of the Technology Advisory Group for Unique Projects'. Ministry of Finance, Government of India.
http://finmin.nic.in/reports/tagup_report.pdf
12. 26 December 2014. 'GST: Gujarat wants 1% extra duty to continue till states want it'. *Economic Times*.

http://articles.economictimes.indiatimes.com/2014-12-26/news/57420384_1_gst-council-anti-dumping-duty-gst-regime

13. July 2010. 'The IT Strategy for GST (Version 0.85)'. Empowered Group on IT Infrastructure on GST headed by Nandan Nilekani (Ministry of Finance, Government of India).
http://finmin.nic.in/GST/IT_Strategy_for_GST_ver0.85.pdf
14. 20 December 2014. 'GST: Finance Minister Arun Jaitley introduces "biggest tax reform" Bill'. *Indian Express.*
http://indianexpress.com/article/india/india-others/goods-service-tax-bill-introduced-in-lok-sabha/

7. Frictionless highways for economic growth

1. Lakshmi, Rama. 23 March 2014. 'India's quest to build modern toll roads hits a pothole'. *Washington Post.*
http://www.washingtonpost.com/world/indias-quest-to-build-modern-toll-roads-hits-a-pothole/2014/03/22/ecfd7b0a-a5fa-11e3-b865-38b254d92063_story.html
2. 26 September 2011. 'Police say toll plaza murder solved; suspected killer arrested'. NDTV.
http://www.ndtv.com/article/india/police-say-toll-plaza-murder-solved-suspected-killer-arrested-136203
3. Bhan, Rohit. 17 October 2012. 'MP who pulled out gun at toll plaza refuses to apologise'. NDTV.
http://www.ndtv.com/article/india/mp-who-pulled-out-gun-at-toll-plaza-refuses-to-apologise-280919
4. 28 September 2013. 'India's informal economy. Hidden value'. *Economist.*
http://www.economist.com/news/asia/21586891-activities-out-sticks-may-add-more-gdp-was-thought-hidden-value
5. June 2014. 'Assessment of Operate-Maintain-Transfer (OMT) and Toll Collection Market for Road Projects in India'. Crisil Research.
http://www.crisil.com/pdf/research/Assessment_of_OMT_TollCollection-Roads.pdf
6. Sunil, B.S. 17 December 2012. 'Why India needs an aggressive highways programme'. *Mint.*

http://www.livemint.com/Opinion/b2TvLxkpta7EEkXWaqnnKK/Why-India-needs-an-aggressive-highways-programme.html

7. Kripalani, Manjeet. 11 March 2009. 'Fabindia Weaves in Artisan Shareholders'. *Bloomberg*.
   http://www.bloomberg.com/bw/stories/2009-03-11/fabindia-weaves-in-artisan-shareholders
   Fabindia website.
   http://www.fabindia.com/company/
8. Kaushik, Manu. 20 March 2011. 'Highways to prosperity'. *Business Today*.
   http://businesstoday.intoday.in/story/toll-road-developers-are-making-the-most-of-indias-ambitious-highway-expansion-plan/1/13611.html
9. Lakshmi, Rama. 21 December 2012. 'An expressway outside New Delhi mirrors India's problems and promise'. *Washington Post*.
   http://www.washingtonpost.com/world/asia_pacific/an-expressway-outside-new-delhi-mirrors-indias-problems-and-promise/2012/12/21/464a6500-3988-11e2-9258-ac7c78d5c680_story.html
10. September 2011. 'Report by Apex Committee for ETC Implementation'. National Highways Authority of India, Government of India.
    http://www.nhai.org/doc/17feb12/etc_apexcomm_report_final_7oct.pdf
    28 June 2010. 'A Unified Electronic Toll Collection Technology for NHAI'. Ministry of Road Transport and Highways, Government of India.
    http://www.nhai.org/ETC%20report.pdf
11. 25 October 2014. 'Nitin Gadkari to Launch Electronic Toll System on Oct 31'. NDTV.
    http://www.ndtv.com/india-news/nitin-gadkari-to-launch-electronic-toll-system-on-oct-31-683790
12. 8 June 2012. 'Shoddy road checkpoint system imposes huge cost on the economy'. *Economic Times*.
    http://articles.economictimes.indiatimes.com/2012-06-08/news/32124221_1_toll-plazas-octroi-and-state-taxes-toll-points
13. 31 October 2014. 'ICICI Bank, Axis Bank tie up for electronic toll collection programme'. *The Hindu Business Line*.
    http://www.thehindubusinessline.com/industry-and-economy/logistics/

icici-bank-axis-bank-tie-up-for-electronic-toll-collection-programme/article6552213.ece

14. 'Congestion Charge'. Transport for London.
http://www.tfl.gov.uk/modes/driving/congestion-charge
'Electronic Road Pricing'. Land Transport Authority, Singapore government.
http://www.lta.gov.sg/content/ltaweb/en/roads-and-motoring/managing-traffic-and-congestion/electronic-road-pricing-erp.html
15. 1 November 2005. 'Road Transport Service Efficiency Study'. World Bank.
http://siteresources.worldbank.org/INTSARREGTOPTRAN SPORT/PublicationsandReports/20747263/Final_version03 NOV2005.pdf

8. Streamlining government spending

1. Cruz, Marcos, and Lazarow, Alexandre. September 2012. 'Innovation in Government: Brazil'. McKinsey & Company.
http://www.mckinsey.com/insights/public_sector/innovation_in_government_brazil
2. 'Where does my money go?'. Open Knowledge Foundation.
http://wheredoesmymoneygo.org
3. 'Open Government Partnership'.
http://www.opengovpartnership.org/
4. Aiyar, Yamini, and Raghunandan, T.R. 'Pull, not push, to open up spending'. *Mint.*
http://www.livemint.com/Opinion/uuVwhsKE8wKNa13sjI2tHN/Pull-not-push-to-open-up-spending.html
5. 31 January 2011. 'Report of the Technology Advisory Group for Unique Projects'. Ministry of Finance, Government of India.
http://finmin.nic.in/reports/tagup_report.pdf
6. 'The Right to Information Act (2005)'. Government of India.
http://rti.gov.in/rti-act.pdf
7. 2012. 'Do schools get their money?' Paisa 2012, Accountability Initiative. ASER.
http://www.accountabilityindia.in/sites/default/files/state-report-cards/paisa_report_2012.pdf

9. Strengthening democracy with technology

1. 18 February 2012. 'Dead people on voter list—Poll panel detects instances of proxy voting, orders repoll'. *Telegraph*.
http://www.telegraphindia.com/1120218/jsp/northeast/story_15148226.jsp#.U9nknFYk_LQ
2. Press Trust of India. 24 April 2014. 'Captains of India Inc vote for stable and decisive govt'. *Economic Times*.
http://articles.economictimes.indiatimes.com/2014-04-24/news/49378102_1_vote-new-government-india-inc
3. Rana, Preetika, and Sugden, Joanna. 15 August 2013. 'India's Record Since Independence'. *Wall Street Journal*.
http://blogs.wsj.com/indiarealtime/2013/08/15/indias-record-since-independence/
4. 21 February 2013. 'Towards Universal Suffrage'. New Zealand Electoral Commission.
http://www.elections.org.nz/right-vote/towards-universal-suffrage
Equal Franchise Act, 1928, UK Parliament.
http://www.parliament.uk/about/living-heritage/transformingsociety/electionsvoting/womenvote/parliamentary-collectionsdelete/equal-franchise-act-1928/
6 August 2015. 'Introduction To Federal Voting Rights Laws'. The United States Department of Justice.
http://www.justice.gov/crt/introduction-federal-voting-rights-laws-1
5. Guha, Ramachandra. 2007. *India after Gandhi: The History of the World's Largest Democracy*. London: Macmillan.
6. 11 March 2014. 'Election Expenditure per elector up by twenty times in 2009 compared to first General Elections'. Press Information Bureau, Election Commission, Government of India.
http://pib.nic.in/newsite/PrintRelease.aspx?relid=104557.
7. Tewari, Saumya. 13 March 2014. 'Betting On Rural Votes This Time Too'. IndiaSpend.
http://www.indiaspend.com/special-reports/betting-on-rural-votes-this-time-too-44937
8. Election Commission of India.
http://eci.nic.in/eci_main1/the_setup.aspx

9. Shekhar, G.C. 7 April 2014. 'Autumn of Al-Seshan. Lest we forget how bad it was till he cleaned it up'. *Telegraph*.
http://www.telegraphindia.com/1140407/jsp/nation/story_18163764.jsp#.U9kJEVYk_LR
10. Ansari, Javed M. 15 December 1994. 'The unsparing rod'. *India Today*.
http://indiatoday.intoday.in/story/cec-t.n.-seshan-tightens-electoral-reform-screws-to-clean-up-entire-election-process/1/294621.html
11. Chouhan, Shashank. 3 April 2014. 'Facts and figures for India's 2014 general election'. Reuters.
http://blogs.reuters.com/india/2014/04/03/facts-and-figures-for-the-2014-general-election
12. Nilekani, Nandan. 2008. *Imagining India: Ideas for the New Century*. New Delhi: Penguin Books.
13. *Electronic Voting Machines, General Elections 2009 Reference Handbook*. Press Information Bureau, Ministry of Information and Broadcasting, Government of India.
http://pib.nic.in/elections2009/volume1/Chap-39.pdf
14. 18 April 2004. 'A Voting Revolution In India?'. *Businessweek*.
http://www.businessweek.com/stories/2004-04-18/a-voting-revolution-in-india
15. Babe, Ann. 4 February 2015. 'Can Open-Source Voting Tech Fix the U.S. Elections System?'. Techonomy.
http://techonomy.com/2015/02/can-open-source-voting-tech-fix-u-s-elections-system
16. In the Supreme Court of India Civil Original Jurisdiction, Writ Petition (Civil) No. 161 of 2004.
http://www.supremecourtofindia.nic.in/outtoday/wp(c)No.161of2004.pdf
17. Press Trust of India. 21 June. 'Election Commission to introduce EVM and VVPAT system for more transparent electronic voting'. *Economic Times*.
http://articles.economictimes.indiatimes.com/2011-06-21/news/29683624_1_polling-stations-voter-electronic-voting-machine
18. Mukul, Akshaya. 1 June 2011. 'EVM with paper trail to be tested in 200 places'. *Times of India*.
http://timesofindia.indiatimes.com/india/EVM-with-paper-trail-to-be-

tested-in-200-places/articleshow/8671700.cms?referral=PM

7 September 2013. 'Nagaland first to use VVPAT system'. *Deccan Herald.* http://www.deccanherald.com/content/355939/nagaland-first-use-vvpat-system.html

19. Press Trust of India. 9 October 2013. 'Supreme Court asks Election Commission to implement paper trail in EVMs'. NDTV. http://www.ndtv.com/india-news/supreme-court-asks-election-commission-to-implement-paper-trail-in-evms-537127

    3 April 2014. 'Bangalore South Chosen for VVPAT System for LS Polls'. *Outlook.* http://www.outlookindia.com/news/article/bangalore-south-chosen-for-vvpat-system-for-ls-polls/835552
20. 2 March 2009. 'The new voting machine: Totalizer'. *Indian Express.* http://archive.indianexpress.com/news/the-new-voting-machine-totalizer/429666/

    11 March 2009. 'No E-Voting In Germany'. European Digital Rights. http://history.edri.org/edri-gram/number7.5/no-evoting-germany
21. Kumar, Vikram. 13 August 2013. 'Delhi hit by massive poll scam: Election commission unearths 13 lakh bogus voters and over 80,000 valid voter cards for dead people'. *Daily Mail.* http://www.dailymail.co.uk/indiahome/indianews/article-2392340/Delhi-hit-massive-poll-scam-Election-commission-unearths-13-lakh-bogus-voters-80-000-valid-voter-cards-dead-people.html
22. Voter turnout data for India, International Institute for Democracy and Electoral Assistance. http://www.idea.int/vt/countryview.cfm?id=105

    28 February 2015. '85 million fake or duplicate names on electoral rolls: EC'. *Hindustan Times.* http://www.hindustantimes.com/india/85-million-fake-or-duplicate-names-on-electoral-rolls-ec/story-gnFsWrjqG64upBMGJ1s7lN.html
23. Kulkarni, Pranav. 18 April 2014. 'An entire society in Kothrud could not vote'. *Indian Express.* http://indianexpress.com/article/cities/pune/an-entire-society-in-kothrud-could-not-vote
24. Press Trust of India. 7 May 2009. 'CEC Navin Chawla nearly missed his vote!'. NDTV.

http://www.ndtv.com/india-news/cec-navin-chawla-nearly-missed-his-vote-393695

25. Chakravarty, Praveen. 28 April 2014. 'The mysteries of "dead" and "deleted" voters'. Rediff.com.
http://www.rediff.com/news/report/slide-show-1-ls-election-the-mysteries-of-dead-and-deleted-voters/20140428.htm#4
26. Gopalaswami, N. 14 December 2012. 'Voting with your fingertips'. *The Hindu*.
http://www.thehindu.com/opinion/lead/voting-with-your-fingertips/article4196670.ece
27. Agarwal, Surabhi, and Aravind, Indulekha. 22 November 2013. 'Voting out errors'. *Business Standard*.
http://www.business-standard.com/article/beyond-business/voting-out-errors-113112201061_1.html
28. 8 April 2012. 'EC, Janaagraha team up to enrol voters'. *DNA India*.
http://www.dnaindia.com/bangalore/report-ec-janaagraha-team-up-to-enrol-voters-1673026
2 February 2015. 'Update the rolls'. *Business Standard*.
http://www.business-standard.com/article/opinion/update-the-rolls-115020201461_1.html
29. 25 January 2015. 'Aadhaar database to be embedded with that of EPIC by 2016: CEC'. *Economic Times*.
http://articles.economictimes.indiatimes.com/2015-01-25/news/58433297_1_aadhaar-database-development-politics-political-parties
15 February 2015. 'ECI to Launch Mission of Electoral Roll Authentication from 1st March'. Press Note, Election Commission of India.
http://eci.nic.in/eci_main1/current/Press_Note_15_02_15-%20Hyderadad-1.pdf
30. National Electoral Roll Purification and Authentication Programme (NERPAP), National Voters' Services Portal.
http://nvsp.in/nerpap.html
31. Ali, Arshad. 22 March 2015. 'Will link Aadhaar number to voter ID cards to ensure no duplicacy: CEC H.S. Brahma'. *Indian Express*.
http://indianexpress.com/article/cities/kolkata/cec-will-link-with-aadhaar-to-ensure-no-duplicate-voter-id/

32. Ferenstein, Gregory. 21 February 2015. 'Watch how quick it is to vote in the most advanced democracy on earth, Estonia'. VentureBeat. http://venturebeat.com/2015/02/21/watch-how-quick-it-is-to-vote-in-the-most-advanced-democracy-on-earth-estonia/
33. Taneja, Mansi, and Agarwal, Surabhi. 23 January 2015. 'E-Voting: Voter IDs to be connected with Aadhaar'. *Business Standard.* http://www.business-standard.com/article/politics/e-voting-voter-ids-to-be-connected-with-aadhaar-115012201274_1.html

10. A new era in politics: The party as a platform

1. Lincoln's plan of campaign in 1840. *Collected Works of Abraham Lincoln.* Vol. 1, p. 181. http://quod.lib.umich.edu/l/lincoln/lincoln1/1:196?rgn=div1;view=fulltext
2. Mehta, Vanya. 4 December 2013. 'Will NRIs' novel campaigning propel Aam Aadmi in today's poll?'. *Deccan Herald.* http://www.deccanherald.com/content/372607/will-nris039-novel-campaigning-propel.html
3. Mail Today Bureau. 7 April 2011. 'Anna Hazare's fight grows via SMSes, Facebook'. *India Today.* http://indiatoday.intoday.in/story/lokpal-bill-anna-hazares-fight-grows-via-smses-facebook/1/134550.html

   Mahesh, Ashwin. 29 December 2013. 'Rise of problem-solving, digital citizen'. *Deccan Chronicle.* http://www.deccanchronicle.com/131229/commentary-sunday-chronicle/article/rise-problem-solving-digital-citizen
4. Ali, Mohammad. 1 October 2013. 'AAP took baby steps on social media, now it's a runaway hit'. *The Hindu.* http://www.thehindu.com/news/cities/Delhi/aap-took-baby-steps-on-social-media-now-its-a-runaway-hit/article5190122.ece

   Ninan, Sevanti. 13 December 2013. 'Learning media strategy from AAP'. *Mint.* http://www.livemint.com/Opinion/HwHIPVrpDJC2Ax0TcTv03N/Learning-media-strategy-from-AAP.html
5. Ghose, Dipankar, and Vatsa, Aditi. 15 February 2015. 'The Big Picture:

What's AAP'. *Indian Express.*
http://indianexpress.com/article/india/india-others/whats-aap/

6. Gupta, Shekhar. 12 February 2015. 'Waking up to a new "usual"'. *India Today*.
http://indiatoday.intoday.in/story/shekhar-gupta-aap-arvind-kejriwal-delhi-election-bjp-congress/1/418585.html
7. Scherer, Michael. 7 November 2012. 'Inside the Secret World of the Data Crunchers Who Helped Obama Win'. *Time.*
http://swampland.time.com/2012/11/07/inside-the-secret-world-of-quants-and-data-crunchers-who-helped-obama-win/
8. Willis, Derek. 25 September 2014. 'Narendra Modi, the Social Media Politician'. *New York Times.*
http://www.nytimes.com/2014/09/26/upshot/narendra-modi-the-social-media-politician.html?smid=tw-nytimes&_r=0&abt=0002&abg=1
9. Bangalore South Lok Sabha constituency, Wikipedia.
https://en.wikipedia.org/wiki/Bangalore_South_(Lok_Sabha_constituency)
10. NGP VAN.
https://www.ngpvan.com

11. The role of government in an innovation economy

1. 10 May 2007. 'To do with the price of fish'. *Economist.*
http://www.economist.com/node/9149142
Iype, George. 13 August 2002. 'Of mobile phones, fishermen and changing lifestyles. Rediff.com.
http://www.rediff.com/news/2002/aug/13spec.htm
2. Jensen, Robert. 2007. 'The Digital Provide: Information (Technology), Market Performance, and Welfare in the South Indian Fisheries Sector'. *The Quarterly Journal of Economics*, 122 (3): 879–924.
http://qje.oxfordjournals.org/content/122/3/879.abstract
3. Srinivasan, Janaki, and Burrell, Jenna. 2013. 'Revisiting the fishers of Kerala, India'. Proceedings of the Sixth International Conference on Information and Communication Technologies and Development: Full Papers, vol. 1, pp. 56–66.
http://dl.acm.org/citation.cfm?id=2516618

4. Reserve Bank of India Notification. 22 August 2014. 'Security issues and risk mitigation measures related to Card Not Present (CNP) transactions'.
https://rbi.org.in/scripts/NotificationUser.aspx?Id=9183&Mode=0
5. Agarwal, Vibhuti, Malhotra, Aditi, and Thoppil, Dhanya Ann. 8 December 2014. 'Uber Banned in Delhi After Rape Allegation'. *Wall Street Journal.*
http://www.wsj.com/articles/uber-banned-in-delhi-after-rape-allegation-1418036309
6. Anand, Jatin. 9 December 2014. 'Verification certificate of driver leaves Delhi Police red-faced'. *The Hindu.*
http://www.thehindu.com/news/cities/Delhi/verification-certificate-of-driver-leaves-delhi-police-redfaced/article6675115.ece?homepage=true.
Sikdar, Shubhomoy. 8 December 2014. 'Uber cab driver had been jailed earlier in a rape case'. *The Hindu.*
http://www.thehindu.com/news/cities/Delhi/uber-cab-driver-had-been-jailed-earlier-in-a-rape-case/article6672130.ece
Press Trust of India. 9 December 2014. 'Uber unsure about driver Shiv Kumar Yadav's police verification'. *Economic Times.*
http://articles.economictimes.indiatimes.com/2014-12-09/news/56879572_1_uber-police-verification-public-service-vehicle
7. Rajagopal, Arjun, Sane, Renuka, and Sundaresan, Somasekhar. 19 December 2014. 'Policy puzzles of the digital nirvana'. Prof. Ajay Shah's blog. (Professor, National Institute of Public Finance and Policy).
http://ajayshahblog.blogspot.in/2014/12/policy-puzzles-of-digital-nirvana.html
8. 9 March 2013. 'All eyes on the sharing economy'. *Economist.*
http://www.economist.com/news/technology-quarterly/21572914-collaborative-consumption-technology-makes-it-easier-people-rent-items
9. 29 December 2014. 'There's an app for that'. *Economist.*
http://www.economist.com/news/briefing/21637355-freelance-workers-available-moments-notice-will-reshape-nature-companies-and
10. Kaka, Noshir, Madgavkar, Anu, Manyika, James, Bughin, Jacques, and Parameswaran, Pradeep. December 2014. 'India's tech opportunity:

Transforming work, empowering people'. McKinsey Global Institute Report.
http://www.mckinsey.com/insights/high_tech_telecoms_ internet/indias_tech_opportunity_transforming_work_empowering_ people

11. Fok, Evelyn. 31 January 2015. 'Sharing economy fails to gain ground in Indian market as startups wary of model'. *Economic Times*.
http://articles.economictimes.indiatimes.com/2015-01-31/news/58650703_1_sharing-economy-taxi-aggregator-vardhan-koshal
12. Mazzucato, M. 2013. *The Entrepreneurial State: Debunking Public vs. Private Sector Myths.* London: Anthem Press.
13. Sally, Madhvi Sally. 30 April 2014. 'Demand for accurate weather predictions heralds burgeoning business opportunity for private forecasting firms. *Economic Times.*
http://articles.economictimes.indiatimes.com/2014-04-30/news/49523457_1_laxman-singh-rathore-skymet-weather-services-weather-data
Srinath, Pavan. 25 July 2014. 'A Citizen Weather Network'.
http://blog.knowyourclimate.org/2014/07/a-citizen-weather-network/
14. Bhosale, Jayashree, and Sally, Madhvi. 4 June 2014. 'Private forecaster Skymet declares monsoon arrival, IMD disagrees'. *Economic Times*.
http://articles.economictimes.indiatimes.com/2014-06-04/news/50330087_1_skymet-monsoon-arrival-imd
15. DARPA Urban Challenge 2005.
http://archive.darpa.mil/grandchallenge05/
16. Fisher, Adam. 18 September 2013. 'Inside Google's Quest To Popularize Self-Driving Cars'. *Popular Science*.
http://www.popsci.com/cars/article/2013-09/google-self-driving-car
Winkler, Rolfe, and Macmillan, Douglas. 2 February, 2015. 'Uber Chases Google in Self-Driving Cars With Carnegie Mellon Deal'. *Wall Street Journal.*
http://blogs.wsj.com/digits/2015/02/02/uber-chases-google-in-self-driving-cars/
Taylor, Edward, and Oreskovic, Alexei. 14 February 2015. 'Apple studies self-driving car, auto industry source says'. Reuters.
http://www.reuters.com/article/2015/02/14/us-apple-autos-idUSKBN0LI0IJ20150214.
17. 18 September 2014. 'Coming to a street near you'. *Economist.*

http://www.economist.com/news/business-and-finance/21618531-making-autonomous-vehicles-reality-coming-street-near-you

18. Pauker, Benjamin. 29 April 2013.'Epiphanies from Chris Anderson'. *Foreign Policy*.
    http://foreignpolicy.com/2013/04/29/epiphanies-from-chris-anderson/
19. Amazon prime air.
    http://www.amazon.com/b?node=8037720011
20. SpaceX.
    http://www.spacex.com
21. Lunar XPRIZE: Team Indus.
    http://lunar.xprize.org/teams/team-indus
22. Wiseman, John. 18 June 2012. Interview with Amory Lovins. Post Carbon Pathways.
    http://www.postcarbonpathways.net.au/wp-content/uploads/2012/10/Amory-Lovins-interview_final.pdf.
23. 1994. *Realizing the Information Future: The Internet and Beyond*. Washington, D.C: The National Academies Press.
    http://www.nap.edu/openbook.php?record_id=4755&page=53
24. Madhu, V. 15 March 2015. 'Why the Prevention of Corruption Act needs to be amended urgently for officers to deliver results'. *Economic Times*.
    http://articles.economictimes.indiatimes.com/2015-03-15/news/60137465_1_public-interest-servant-corruption-act
25. January 2013. 'Report of the Expert Committee on the HR Policy for e-Governance'. Department of Electronics and Information Technology, Ministry of Communications and Information Technology, Government of India.
    https://negp.gov.in/pdfs/HR_Policy_for_eGovernance.pdf

12. Towards a healthy India

1. Anand, Geeta. 'One Family's Futile Quest to Get Help for Their Baby'. 30 July 2011. *Wall Street Journal*.
   http://www.wsj.com/articles/SB1000142405311190480030457647436269 7223954
   Hayden, Michael Edison. 2014. 'In West Bengal, a Pilgrimage of the Sick'. Roads & Kingdoms.

http://roadsandkingdoms.com/2014/in-west-bengal-a-pilgrimage-of-the-sick/

2. Bahri, Charu. 'Diabetes: The Epidemic That Indians Created'. IndiaSpend.
   http://www.indiaspend.com/sectors/health/diabetes-the-epidemic-that-indians-created-18722
3. Sinha, Kounteya. 19 November 2012. '44 lakh Indians don't know they are diabetic'. *Times of India.*
   http://timesofindia.indiatimes.com/india/44-lakh-Indians-dont-know-they-are-diabetic/articleshow/17274366.cms?
   13 May 2013. 'Helping India Combat Persistently High Rates of Malnutrition'. World Bank.
   http://www.worldbank.org/en/news/feature/2013/05/13/helping-india-combat-persistently-high-rates-of-malnutrition
4. Press Trust of India. 9 July 2014. 'Economic Survey 2014: "More needs to be done to give quality, affordable healthcare"'. *Economic Times.*
   http://articles.economictimes.indiatimes.com/2014-07-09/news/51248036_1_health-sector-infant-mortality-rate-national-health-mission
   Kalra, Aditya. 28 February 2015. 'India keeps tight rein on public health spending in 2015–16 budget'. Reuters.
   http://in.reuters.com/article/2015/02/28/india-health-budget-idINKBN0LW0LQ20150228
   India Health Statistics 2014, OECD.
   http://www.oecd.org/els/health-systems/ Briefing-Note-INDIA-2014.pdf.
5. December 2012. 'India Healthcare: Inspiring Possibilities, Challenging Journey'. McKinsey and Co.
   http://www.mckinsey.com/global_locations/asia/india/en/latest_thinking
6. December 2014. 'National Health Policy 2015 Draft'. Ministry of Health and Family Welfare, Government of India.
   http://www.mohfw.nic.in/showfile. php?lid=3014.
7. Lalmalsawma, David. 4 June 2014. 'Telemedicine in India Might Be Just What the Doctor Ordered'. Reuters India.
   http://blogs.reuters.com/india/2014/06/04/telemedicine-in-india-might-be-just-what-the-doctor-ordered/

8. Tran, Joseph et al. 2012. 'Smartphone-based Glucose Monitors and Applications in the Management of Diabetes: An Overview of 10 Salient "Apps" and a Novel Smartphone-connected Blood Glucose Monitor'. *Clinical Diabetes*, vol. 30 (4), pp. 173–178.
http://clinical.diabetesjournals. org/content/30/4/173.full.
Hoskins, Mike. 28 March 2013. 'Dario: Turning Your Smartphone Into a Glucose Meter'. Healthline.
http://www.healthline.com/diabetesmine/the-dario-turns-your-smartphone-into-a-meter.
Aungst, Timothy. 6 November 2013. 'The next frontier, attaching health sensors directly into your smartphone'. iMedicalApps.
http://www.imedicalapps.com/2013/11/frontier-health-sensors-smartphone/
9. 2011. 'Cardiovascular Diseases in India: Challenges and Way Ahead'. Deloitte.
http://www2.deloitte.com/content/dam/Deloitte/in/Documents/life-sciences-health-care/in-lshc-cardio-noexp.pdf
10. Gawande, Atul. 24 January 2011. 'The Hot Spotters'. *The New Yorker*.
http://www.newyorker.com/magazine/2011/01/24/the-hot-spotters
11. 'HIT or Miss. The digitisation of medical records is getting closer'. 16 April 2009. *Economist*.
http://www.economist.com/node/13438006.
12. Chaudhury, Nazmul et al. 2006. 'Missing in Action: Teacher and Health Worker Absence in Developing Countries'. *Journal of Economic Perspectives*, vol. 20 (1): 91–116.
http://pubs.aeaweb.org/doi/pdfplus/10.1257/089533006776526058
Muralidharan, K. et al. 2011. 'Is There a Doctor in the House? Medical Worker Absence in India'. Working paper, Harvard University.
http://scholar.harvard.edu/kremer/publications/there-Doctor-House-Medical-Worker-Absence-India

13. Teaching the next generation

1. Salve, Prachi. 2 May 2015. '$94 Billion on Basic Education Doesn't Address Teaching Crisis'. IndiaSpend.
http://www.indiaspend.com/cover-story/94-billion-on-basic-education-doesnt-address-teaching-crisis-46964

2. Tewari, Saumya. 19 May 2015. 'Teachers Get 80% of Education Expenditure: New Report'. IndiaSpend.
http://www.indiaspend.com/cover-story/teachers-get-80-of-education-expenditure-new-report-31884
3. CensusInfo India 2011, Final Population Totals.
http://censusindia.gov.in/2011census/censusinfodashboard/index.html
4. Banerjee, Rukmini. 19 January 2012. 'The Crisis in Learning'. *Indian Express*.
http://indianexpress.com/article/opinion/columns/the-crisis-in-learning/99/
5. '80 million kids drop out without completing basic schooling: UNICEF'. 12 April 2013. *The Hindu*.
http://www.thehindu.com/todays-paper/tp-national/80-million-kids-drop-out-without-competing-basic-schooling-unicef/article4608635.ece
6. 'The National Picture'. Annual Status of Education Report, 2013.
http://img.asercentre.org/docs/Publications/ASER%20Reports/ASER_2013/ASER2013_report%20sections/thenationalpicture.pdf
7. Chavan, Madhav. 2013. 'Old Challenges for a New Generation'.
http://www.pratham.org/templates/pratham/images/madhavchavanarticle.pdf
8. Muralidharan, Karthik. 18 March 2013. 'Using Evidence for Better Policy: The Case of Primary Education in India'.
http://www.ideasforindia.in/article.aspx?article_id=119
9. Khan, Salman. 9 April 2011. 'Turning the Classroom Upside Down'. *Wall Street Journal*.
http://www.wsj.com/articles/SB1000142405274870410160457624871342074788 4
10. Muralidharan, Karthik. 25 November 2013. 'An Evidence-based Proposal to Achieve Universal Quality Primary Education in India'.
http://ideasforindia.in/article.aspx?article_id=222
11. Phadnis, Shilpa. 18 November 2014. 'Nandan Nilekani now logs in to boost primary education'. *Times of India*.
http://timesofindia.indiatimes.com/india/Nandan-Nilekani-now-logs-in-to-boost-primary-education/articleshow/45184536.cms?
12. Patel, Nilay. 22 January 2015. 'Bill Gates is guest-editing The Verge in February. Technology will build a better, safer, healthier world by 2030'. *The Verge*.

http://www.theverge.com/2015/1/22/7870497/bill-gates-interview-future-verge-guest-editor.

13. Nair, Malini. 18 August 2013. 'MOOCs click with Indians'. *Times of India.*
    http://timesofindia.indiatimes.com/home/sunday-times/deep-focus/MOOCs-click-with-Indians/articleshow/21890105.cms
14. Sakshat.
    http://www.sakshat.ac.in
    NPTEL.
    http://nptel.ac.in
15. '47% of graduates of 2013 unemployable for any job: Study'. 27 December 2013. *Economic Times.*
    http://articles.economictimes.indiatimes.com/2013-12-27/news/45626762_1_employability-aspiring-minds-graduates
16. Voucher schemes in India, School Choice, Centre for Civil Society.
    http://www.schoolchoice.in/voucherschemeindia.php
17. Hebbar, Nistula. 29 August 2014. 'PM Modi's big plan: Get education, medical & birth records online in a Digital Locker'. *Economic Times.*
    http://articles.economictimes.indiatimes.com/2014-08-29/news/53362935_1_prime-minister-narendra-modi-suggestions-government-offices
18. Nandakumar, Indu, and Julka, Harsimran. 8 August 2012. 'Resume fraud weighs on Indian IT industry, one in five CVs contain fake information'. *Economic Times.*
    http://articles.economictimes.indiatimes.com/2012-08-08/news/33100650_1_nasscom-fraud-national-skills-registry
19. Madhusoodan, M.K., and Phadnis, Shilpa. 29 September 2013. 'Fake companies help fluff CVs for a fee'. *Times of India.*
    http://timesofindia.indiatimes.com/india/Fake-companies-help-fluff-CVs-for-a-fee/articleshow/23227973.cms
20. Sen, Anirban. 14 July 2015. 'The Return of Nandan Nilekani'. ETtech.
    http://tech.economictimes.indiatimes.com/news/people/the-return-of-nandan-nilekani/48059846
21. 12 January 2010. 'Shri Kapil Sibal announces new initiative to demat degrees: National database of academic qualifications in electronic

format envisaged'. Press Information Bureau, Government of India. http://pib.nic.in/newsite/erelease.aspx?relid=56816.

'Cabinet clears Academic Depository Bill'. 23 March 2011. *The Hindu*. http://www.thehindu.com/news/national/article1561573.ece?mstac=0

14. Switching on our power sector

1. Dedhia, Sonil. '*Katiyabaaz* is electrifyingly real'. Rediff.com. http://www.rediff.com/movies/report/review-katiyabaaz-is-electrifyingly-real/20140822.htm.
2. Energy Statistics 2015, Central Statistics Office, National Statistical Organisation, Ministry of Statistics and Programme Implementation, Government of India. http://mospi.nic.in/Mospi_New/upload/Energy_stats_2015_26mar15.pdf
3. 26 June 2014. 'India energy data and analysis', U.S. Energy Information Administration. http://www.eia.gov/beta/international/analysis.cfm?iso=IND
4. 1 August 2012. 'Power grid failure: 30% power lost to theft, politics. *Times of India*. http://timesofindia.indiatimes.com/india/Power-grid-failure-30-power-lost-to-theft-politics/articleshow/15300572.cms
5. McCarthy, Julie. 17 February 2015. 'What's it like to live without electricity? Ask an Indian villager'. npr.org. http://www.npr.org/sections/goatsandsoda/2015/02/17/386876116/whats-it-like-to-live-without-electricity-ask-an-indian-villager
6. 24 June 2014. 'Switching on power sector reform in India'. World Bank. http://www.worldbank.org/en/news/feature/2014/06/24/switching-on-power-sector-reform-in-india
7. June 2014. 'Sustainable Energy for All: Compendium of the Work of UN Agencies in India'. http://www.un.org.in/img/uploads/UNIDO-Compendium-Report-27-05-14.pdf
8. 'Building the smart grid'. 4 June 2009. *Economist*. http://www.economist.com/node/13725843
9. 'Estimating the Costs and Benefits of the Smart Grid'. 2011. Technical

Report, Electric Power Research Institute.
http://www.rmi.org/Content/Files/EstimatingCostsSmartGRid.pdf

10. Sharma, Amol, Chaturvedi, Saurabh, and Choudhury, Santanu. 31 July 2012. 'India's Power Network Breaks Down'.
http://www.wsj.com/articles/SB10000872396390444058045775604131 78678898
11. McKerracher, C., and Torriti, J. 2013. 'Energy consumption feedback in perspective: Integrating Australian data to meta-analyses on in-home displays'. *Energy Efficiency*, vol. 6 (2), pp. 387–405.
12. Scott, Mark. 16 November 2009. 'How Italy beat the world to a smarter grid'. *Businessweek*.
http://www.businessweek.com/globalbiz/content/nov2009/gb20091116_319929.htm
13. 15 January 2015. 'Invisible fuel'. *Economist*.
http://www.economist.com/news/special-report/21639016-biggest-innovation-energy-go-without-invisible-fuel
14. 1 March 2014. 'Negawatt Hour'. *Economist*.
http://www.economist.com/news/business/21597922-energy-conservation-business-booming-negawatt-hour
15. 15 January 2015. 'Renewables. We make our own'. *Economist*.
http://www.economist.com/news/special-report/21639020-renewables-are-no-longer-fad-fact-life-supercharged-advances-power
16. Dzieza, Josh. 13 February 2015. 'Why Tesla's battery for your home should terrify utilities'. *The Verge*.
http://www.theverge.com/2015/2/13/8033691/why-teslas-battery-for-your-home-should-terrify-utilities
17. Ramesh, Jairam. 24 February 2015. 'Gigawatts-plus approach'. *Mint*.
http://www.livemint.com/Opinion/MvWm203A52IBxf3iN72QUM/Gigawattsplus-approach.html
18. IT task force report for power sector.
http://www.infosys.com/newsroom/features/Documents/2002-ITTask-Force-Report.pdf
2008. 'Technology: Enabling the transformation of power distribution'. Report by CSTEP and Infosys.
http://www.infosys.com/newsroom/features/Documents/power-sector-report.pdf

19. Cardwell, Diane. 18 April 2015. 'Solar power battle puts Hawaii at forefront of worldwide changes'. *New York Times*.
http://www.nytimes.com/2015/04/19/business/energy-environment/solar-power-battle-puts-hawaii-at-forefront-of-worldwide-changes.html?_r=1

15. Justice delayed is justice denied

1. 9 May 2013. 'Only 13 judges for every ten lakh people in India'. *The Economic Times*.
http://articles.economictimes.indiatimes.com/2013-05-09/news/39144068_1_judges-high-courts-justice
2. 1 December 2014. 'More than 3 crore court cases pending across country'. NDTV.
http://www.ndtv.com/india-news/more-than-3-crore-court-cases-pending-across-country-709595
3. 6 March 2010. 'Courts will take 320 years to clear backlog cases: Justice Rao'. *The Times of India*.
http://timesofindia.indiatimes.com/india/Courts-will-take-320-years-to-clear-backlog-cases-Justice-Rao/articleshow/5651782.cms?referral=PM
4. 13 February 2009. 'It would take Delhi HC 466 yrs to clear backlog: CJ'. *The Indian Express*.
http://archive.indianexpress.com/news/it-would-take-delhi-hc-466-yrs-to-clear-backlog-cj/423127/
Garg, Abhinav.' 30 May 2012. 'At 5 minutes per case, Delhi high court clears 94,000 in 2 years'. *Times of India*.
http://timesofindia.indiatimes.com/city/delhi/At-5-minutes-per-case-Delhi-high-court-clears-94000-in-2-years/articleshow/13663493.cms?referral=PM
5. Krishnan, K., Jayanth, Raj Kumar, C. 2011. 'Delay in Process, Denial of Justice: The Jurisprudence and Empirics of Speedy Trials in Comparative Perspectives'. *Georgetown Journal of International Law*, vol. 42. Jindal Global Legal Research Paper No. 5/2011.
17 January 2013. 'India to have 15 crore pending cases by 2040, report says'. *The Times of India*.

http://timesofindia.indiatimes.com/india/India-to-have-15-crore-pending-cases-by-2040-report-says/articleshow/18054608.cms?referral=PM

6. Adhivarahan, V. 3–4 May. 2003. 'Case Management and ADR for Banking Sector'. Paper presented in International Conference on ADR and Case Management. Law Commission of India, Ministry of Law and Justice, Government of India.
   http://lawcommissionofindia.nic.in/adr_conf/adivarahan1.pdf.
7. Hamilton, Alexander. The Federalist No. 78.
   http://www.constitution.org/fed/federa78.htm
8. 9 April 2007. 'CJI: Deciding validity of law part of judicial function'. *The Hindu*.
   http://www.thehindu.com/todays-paper/tp-national/cji-deciding-validity-of-law-part-of-judicial-function/article1825458.ece
9. '1993 Mumbai blasts trial longest in India's history'. Rediff.com.
   http://www.rediff.com/news/report/mumbai-blasts-1993-trial-longest-in-indias-history/20130321.htm
10. 24 June 2011. 'Reforms could see disposal of cases in three years'. *The Hindu*.
    http://www.thehindu.com/news/national/reforms-could-see-disposal-of-cases-in-three-years/article2129739.ece
11. e-Courts: A Mission Mode Project to Transform Justice Delivery by ICT Enablement of Courts.
    http://ecourts.gov.in
12. Judicial Administration, United States Courts.
    http://www.uscourts.gov/about-federal-courts/judicial-administration

16. Rebooting India: Realizing a billion aspirations

1. Kennedy, Robert, F. 18 March 1968. Speech at the University of Kansas.
   http://www.jfklibrary.org/Research/Research-Aids/Ready-Reference/RFK-Speeches/Remarks-of-Robert-F-Kennedy-at-the-University-of-Kansas-March-18-1968.aspx
2. 2015. 'Long-term macroeconomic forecasts: Key trends to 2050'. A special report from The Economist Intelligence Unit.

http://pages.eiu.com/rs/783-XMC-194/images/Long-termMacroeconomicForecasts_KeyTrends.pdf

3. 'Informal sector and conditions of employment in India'. NSS 68th Round (July 2011–June 2012). National Sample Survey Office, Ministry of Statistics and Programme Implementation, Government of India.

   http://mospi.nic.in/Mospi_New/upload/nss_report_557_26aug14.pdf

4. Economic Survey 2104–15: 'The Fourteenth Finance Commission (FFC)—Implications for Fiscal Federalism in India?'. Ministry of Finance, Government of India.

   http://indiabudget.nic.in/es2014-15/echapvol1-10.pdf

INDEX